Geocultural Power

THE SILK ROADS SERIES
James A. Millward, Series Editor

The Silk Roads series is made possible by the generous support of the Henry Luce Foundation's Asia Program. Founded in 1936, the Luce Foundation is a not-for-profit philanthropic organization devoted to promoting innovation in academic, policy, religious, and art communities. The Asia Program aims to foster cultural and intellectual exchange between the United States and the countries of East and Southeast Asia, and to create scholarly and public resources for improved understanding of Asia in the United States.

Islam and World History: The Ventures of Marshall Hodgson
Edited by Edmund Burke III and Robert J. Mankin
Published 2018

Sacred Mandates: Asian International Relations since Chinggis Khan
Edited by Timothy Brook, Michael van Walt van Praag, and Miek Boltjes
Published 2018

Geocultural Power

China's Quest to Revive the Silk Roads for the Twenty-First Century

TIM WINTER

The University of Chicago Press   Chicago and London

The University of Chicago Press, Chicago 60637
The University of Chicago Press, Ltd., London

For more information, contact the University of Chicago Press, 1427 E. 60th St., Chicago, IL 60637.
Published 2019
Printed in the United States of America

28 27 26 25 24 23 22 21 20 19 1 2 3 4 5

ISBN-13: 978-0-226-65821-6 (cloth)
ISBN-13: 978-0-226-65835-3 (paper)
ISBN-13: 978-0-226-65849-0 (e-book)
DOI: https://doi.org/10.7208/chicago/9780226658490.001.0001

Any additional permissions/subsidy info needed—to come from acquisitions/contracts/sub rights

Library of Congress Cataloging-in-Publication Data

Names: Winter, Tim, 1971– author.
Title: Geocultural power : China's quest to revive the Silk Roads for the twenty-first century / Tim Winter.
Description: Chicago : The University of Chicago Press, 2019. | Series: Silk roads | Includes bibliographical references and index.
Identifiers: LCCN 2019005356 | ISBN 9780226658216 (cloth : alk. paper) | ISBN 9780226658353 (pbk. : alk. paper) | ISBN 9780226658490 (e-book)
Subjects: LCSH: China—Foreign relations. | Asia—History—Political aspects. | Silk Road—Political aspects. | History—Political aspects.
Classification: LCC DS779.47.W56 2019 | DDC 337.51—dc23
LC record available at https://lccn.loc.gov/2019005356

♾ This paper meets the requirements of ANSI/NISO Z39.48-1992 (Permanence of Paper).

To a Trojan Soul

Contents

Illustrations

Preface

In late 2016 it seemed as though the world was united by infrastructure, or, to be more precise, its possibilities for future making. But Donald Trump's grand proclamations to "make America great again" by rebuilding roads, airports, and energy plants was only half the story. At a time when YouTube clips, news channels, and online newspapers left audiences around the world transfixed, if not a little amused and terrified, few people in the United States, Canada, Australia, or Europe were aware of the other infrastructure story being covered by news outlets on a near-daily basis in Asia, Africa, and the Middle East. BBC World appeared to have missed the news about the Belt and Road Initiative, headlining the Beijing summit in 2017 as the project's "launch." For those who had followed Belt and Road since its actual launch in 2013, recurring questions surrounded how, whether, and at what speed this extraordinarily ambitious project would evolve. In contrast to Donald Trump's isolationist rhetoric of domestic infrastructure, Xi Jinping's ambitions are to defy borders and integrate Eurasia via telecommunications, finance, legislation, and tourism, and by road, rail, sea, and pipeline. Although a healthy skepticism is required to cut through the hyperbole that has surrounded Belt and Road, events through 2018 and early 2019 offered clear signposts to a world changing in historically significant ways. In Washington the funding of a wall became the metaphor for intensely contested visions of a country's future, leading to a level of paralysis in government outdone only by London. Around the same time, in Beijing preparations were being made for the second Belt and Road summit, with Vladimir Putin among the heads of state invited to discuss plans and progress in continental connectivity. This book speaks to such changes, but it does so by considering how Belt and Road *makes history* in quite different ways: how, on the one hand, it seeks to reshape the international trade and political relations of today, but, on the other hand, how it rewrites history, digs up the past, and reworks it for purposes of expediency. Very different questions thus sit alongside one another here as part of an attempt to make sense of some trends and shifts I have been observing across Asia over some time. The "rise" of China has indeed been extraordinary. One of my hopes here is to complement the many theses on its growing economic, technological, and military might by exploring the question of whether China is also a rising cultural power, and, if so, in which ways are others responding as it begins to yield that strength?

As the chapters here demonstrate, Belt and Road is an undertaking in connectivity. The vast majority of the analysis to date has scrutinized its economic, political, and infrastructure linkages. Little attention has been given to Belt and Road as a site of cultural production and cultural politics. But when I took up such themes, it soon became apparent that the mantra of connectivity demanded a particular approach, one that involved crossing geographical boundaries and gingerly jumping over disciplinary fences. Specialists in certain fields will no doubt feel I caught my trousers on the wire, as they will be able to furnish the questions and issues raised here with considerably more detail and trace many further analytical connections. But in the process I have been struck by the questions Belt and Road raises about the merits and pitfalls of crossing frontiers in the quest to understand its multitudinous implications. Grasping at Belt and Road has been a fascinating and highly challenging task. The chapters that follow do not proclaim to tell the whole story; rather, they highlight connections, trends, and assemblages that I believe will continue to consolidate over the medium term, albeit in directions and ways that are impossible to fully anticipate. Six years after launch, the political commitment to the Belt and Road Initiative remains as strong as ever. Xi Jinping's reputation appears tied to its long-term success, and the rapid geopolitical and economic shifts occurring across Asia-Pacific mean governments and businesses cannot afford to ignore its gravitational pull. India is among those countries that remain skeptical—indeed, suspicious—of Beijing's agenda. It would be foolish to think regional tensions in Asia will be resolved and smoothed out by a vision of harmony and win-win cooperation led by one rising power. The region's history tells us otherwise. But as a political project of historical imagination, Belt and Road represents a fascinating attempt to engineer particular futures into existence. In mapping out some of the interconnected pathways of this, I also hope to cast a different light on how the past is used in present-day politics and in the production of particular forms of social and spatial governance. Countries have long cooperated to make and remake history, and in so doing, they have co-opted and pacified the past to redefine social and national boundaries. Belt and Road draws us into these worlds on an unprecedented scale.

There are a number of people I would like to thank for their assistance in making this volume possible. First, working with the University of Chicago Press has been a real pleasure. From the outset, Priya Nelson has been a wonderful editor, engaged and supportive throughout. Susan Karani, Kristen Raddatz, and Dylan Montanari have also been extremely helpful in guiding the project forward through to completion. Particular thanks go to Katherine Faydash for her extraordinary eye for detail and flow. The text benefited greatly from her copyediting talents and care. I also greatly appreciate the time spent by James Millward reading and commenting on the manuscript, and for endorsing the book at those moments when it mattered. I am extremely grateful to the guidance offered by the University of Chicago Press's anonymous reviewers. They challenged, encouraged, and critiqued in equal measure, and in ways that opened up whole new directions of analysis.

The work has benefited from a number of archives and libraries. In Paris, Adele Torrance and Eng Sengsavang at UNESCO have been very helpful over a number of years. The Museum of Islamic Arts Library in Doha proved a valuable source for material on Silk Road histories. For the supply of images I thank Antik Bar, the British Library, and GRAD London. The evident trend toward dwindling resources for HASS research is a worrying one. I am therefore very grateful to the support provided by the Australia Research Council, and this book is among the outcomes of two grants, DP140102991 (Crisis in World Heritage) and FT170100084 (Heritage diplomacy and One Belt One Road). I also thank the Asia Research Centre, Singapore, for a fellowship that allowed significant developments to be made in the theoretical framing of the text. In particular I thank Prasenjit Duara for his support and interest. Young-pil Kwon went at great lengths to send through otherwise unavailable publications. I am indebted to Kwa Chong Guan for his insights into the Maritime Silk Road and strong enthusiasm for the project and the connections being traced. Roland Lin Chih-Hung offered guidance on UNESCO matters, and Tim Williams very kindly shared reports and studies that helped open up my understanding of a complex picture of Silk Road world heritage. Melathi Saldin helped with clarifying Belt and Road Initiative details in Sri Lanka and kindly tracked down Zheng He.

In the process of writing and thinking—in the comprehending of where Belt and Road actually takes us—I have benefited from the thoughts of colleagues and friends across a number of countries. In no particular order I thank Wu Zongjie, Yang Jianping, Michael Keane, Ed Wastnidge, Brett Bennett, Ali Mozaffari, Minoo Sinaiee, Ien Ang, James Liebold, Baogang He, Emma Mawdsley, Rodney Harrison, Jorge Otero-Pailos, Thalia Kennedy, Adèle Esposito, Zhao Taotao, Hyung il Pai, Luke James, Judy and Mark, Ali and Toby, Daniel Schumacher, and Dougald O'Reilly. Kristal Buckley, Erin Linn-Tynen Edward Vickers, Mark Ravinder Frost, Jiat Hwee Chang, and Kearrin Sims have also been part of such conversations, but they are among those who have very kindly given up precious time to read part or all of the manuscript. Writing this on a beautiful handmade desk, gifted to a son on the other side of the world, was a daily reminder of unwavering belief and support. Over the years, the belt has loosened and time has been lost down dead-end roads, but despite this, Wantanee has given her love and support. Finally, I would like to offer particular thanks to Toyah Horman. This book would not have been possible without Toyah's talents and willingness to dig further, go back and check, meticulously plot, gather and arrange, and ride the roller coaster. Toyah helped make this the most fascinating and enjoyable project I have worked on, and I am very grateful for that.

Work Together for a Bright Future of China-Iran Relations

XI JINPING, JANUARY 2016

As I am about to embark on my state visit to Iran at the invitation of President Rouhani, I am looking forward to in-depth exchange of views on deepening China-Iran relations in the new era as well as on major international and regional issues and working together with my Iranian hosts to bring the relationship to a new stage.[1]

This will be my first trip to Iran, yet like many other Chinese, I do not feel like a stranger in your ancient and beautiful country, thanks to the Silk Road that linked our two great nations for centuries and to the many legendary stories recorded in history books of our friendly exchanges.

Over 2,000 years ago, during the West Han dynasty in China, the Chinese envoy Zhang Qian's deputy came to Iran and received a warm welcome. Seven centuries later during the Tang and Song dynasties, many Iranians came to China's Xi'an and Guangzhou to study, practice medicine and do business. In the 13th Century, the famous Iranian poet Saadi wrote about his unforgettable travel to Kashgar, Xinjiang. In the 15th Century, a renowned Chinese navigator Zheng He from the Ming Dynasty led seven maritime expeditions, which took him to Hormuz in southern Iran three times.

The much-prized Persian carpet is weaved out of a fusion of China's silk and Iran's sophisticated techniques. And the exquisite blue and white porcelain is produced thanks to a mixture of Iran's "smaltum" (a type of material containing cobalt, unique to Iran) and China's advanced skills. Via Iran, China's lacquerware, pottery, as well as papermaking, metallurgical, printing, and gunpowder making skills were spread to the west end of Asia, and further on to Europe. And from Iran and Europe, pomegranate, grape, olive, as well as glass, gold and silver ware[s] were introduced into China.

It almost seemed that our two countries were just a camel ride or a boat trip away from each other. Indeed, the thousand-mile-long land and maritime silk roads made it possible for two ancient civilizations and peoples to embrace and befriend each other. As Saadi wrote, those that are far away and are of times long past deserve to be cherished more.

In history, China and Iran made important contributions to opening the Silk Road and promoting exchanges between Eastern and Western civilizations. The China-Iran friendly exchanges in the 45 years of our diplomatic relations have continued to embody the Silk Road spirit of peace, cooperation, openness, inclusiveness, mutual learning and mutual benefit.

Since the inception of our diplomatic ties in 1971, the China-Iran relationship has stood the test of international changes and maintained a momentum of sound and steady development. We have given each other mutual understanding and mutual trust in good times and bad. . . . Our bilateral trade jumped from tens of millions of US dollars in the 1970s to $51.8 billion in 2014, and China has stayed Iran's biggest trading partner for six years in a row. Our two countries have also enjoyed very close people-to-people and cultural ties. As a Chinese saying goes, good friends feel close even when they are thousands of miles apart. The friendship between our peoples has become a significant driver of the friendly relations between our two countries. . . . In 2013, I put forward the proposal of jointly building the Silk Road Economic Belt and the 21st Century Maritime Silk Road, which received positive response from Iran. As two important stops on the ancient Silk Road, both China and Iran have high expectations for reviving this road of peace, friendship, and cooperation.

Cooperation between China and Iran under the framework of the Belt and Road Initiative may focus on the following areas:

Enhancing political mutual trust to cement the foundation for cooperation. Mutual trust ensures success while distrust spells failure. This is true for state-to-state relations as well as interpersonal relationships. Countries along the ancient Silk Road have built trust, deepened friendship and enhanced cooperation through their exchanges stretching over 2,000 years. Over the past 45 years of diplomatic relations, China and Iran have enhanced traditional friendship and achieved fruitful results in practical cooperation despite difficulties and obstacles. Today, we need all the more to build on this positive spirit to step up policy communication, accommodate each other's concerns, build more consensus and lay a more solid foundation for our cooperation. We will establish a comprehensive strategic partnership and increase exchanges between political parties, legislatures and at the sub-national levels.

Pursuing win-win outcomes and common prosperity. . . . China will work to synergize its development strategy with that of Iran and deepen win-win cooperation to fully exert respective strengths for the benefit of our peoples and to achieve greater common prosperity.

Promoting connectivity and expanding practical cooperation. Connectivity is the artery of the Belt and Road Initiative. In building connectivity, we should give priority to Asian countries and start with transportation infrastructure. China has a strong competitive edge in areas such as railway, electricity, telecommunications, mechanical engineering, metallurgy and construction materials. We committed [US]$40 billion to the establishment of a Silk Road Fund in 2014 to support relevant cooperation projects in countries along the Belt and Road. Iran is strategically located and has distinctive geographical advantages. China is ready to deepen cooperation with Iran on building roads, railway, sea routes and the Internet, and facilitating East-West connectivity in Asia and on this basis better promote trade and investment liberalization among countries along the Belt and Road, lower the cost of the cross-border movement of people, goods, and capital, and expand energy, resources, and industrial cooperation.

Upholding openness and inclusiveness and encouraging inter-civilization exchange. The Chinese people often say the value of friendship lies in heart-to-heart communication. A Persian proverb also goes, there is telepathy between hearts. Different countries, nations, and civilizations should carry out exchanges and mutual learning and live in harmony with each other. Both the Chinese and Iranian cultures have unique strengths and the two peoples have benefited from mutual learning for centuries. We need to step up exchanges in culture, education, information, publishing, tourism and other fields and encourage more exchanges between the youth and students, so that the spirit of the Silk Road will be passed on from generation to generation and our peoples will develop enduring bonds.

Pomegranate is well liked in China for its crimson flower and bountiful seeds, for which it came to symbolize plentifulness and prosperity. Introduced from Iran to China centuries ago, the fruit bears witness to the history of friendly exchanges between the Chinese and Iranians along the Silk Road and augurs even more fruitful cooperation between our two countries.

The long distance between Beijing and Tehran is no obstacle to the interaction or cooperation between China and Iran, nor to the friendship and exchanges between our peoples. China is ready to join hands with Iran to renew the Silk Road spirit and create an ever better future for China-Iran relations.

ONE

From Camels and Sails to Highways and Refineries

History is like the tide of water that always rolls forward. The peaceful development, people's well-being, openness and inclusiveness, win-win cooperation are the irresistible historical trend.

SUN WEIDONG, CHINESE AMBASSADOR TO PAKISTAN, NOVEMBER 25, 2015[1]

In broad terms, this book considers the use of history and heritage for political ends. In that respect, it both builds on and significantly departs from a line of academic inquiry that has blossomed across a number of disciplines in recent decades. Two themes have dominated this space, nationalism and conflict. Numerous studies have been conducted into the various ways in which architecture and archaeological landscapes have been deployed in the politics of nation building. We have seen why states both build monuments and monumentalize existing structures to commemorate, fix, and impose preferred narratives of history. As Benedict Anderson lucidly demonstrated, the preservation and symbolic loading of the material past as heritage has been pivotal in creating the "imagined communities" of nation-states in the modern era.[2] And with the language of heritage deeply entwined with notions of identity, territory, and belonging, a vast literature has addressed the different ways in which the past has been a source of contestation and violence. The notion of history wars has become familiar in accounting for the competing claims that states and other groups forcibly make over the past and its material culture.[3] Conflicts in the Middle

East and southeastern Europe have demonstrated how cultural heritage can be explicitly targeted, whereby the destruction of churches, mosques, or entire cities has been incorporated into military strategies designed to undermine the religious, cultural, or territorial bonds of populations. In March 2001 a remote valley in central Afghanistan became the focal point of global media attention, with the dynamiting of two Buddha statues carved out of the sandstone cliffs, measuring 115 and 174 feet, respectively. In smashing idols, the Taliban claimed their destruction was in accordance with Islamic law. In 2013, the Islamic State (IS) began a sustained program of heritage destruction and looting across multiple locations in Syria and Iraq. The premise for this was twofold. Economically, the smuggling of looted antiquities provided IS with a significant stream of revenue. Politically, the deliberate destruction of pre-Islamic archaeological sites, churches, and Shiite shrines and mosques was intended to eradicate difference and create a new state based on a singular narrative of history. Libraries were also targeted, with documents and books burned in a process Robert Bevan has documented as the intentional "destruction of memory."[4] With the smashing of heritage sites in Mali by Ahmad al-Faqi al-Mahdi in 2012, and the subsequent investigation and indictment by the International Criminal Court, we have also entered an era where such acts are recognized within international law as a war crime.

Not surprisingly, this seeming increase in the politicization of history and its cultural artifacts has been the subject of intense debate and consternation among scholars and heritage agencies around the world. Countless conferences, reports, dissertations, and scholarly publications have been dedicated to understanding the motivations for destruction and to documenting its scale. Intense debate has surrounded whether the heritage destruction by IS, publicized via a new age of global web-based broadcasting, represents something fundamentally new, or whether it merely serves as the latest manifestation of a tradition of iconoclasm and cultural erasure stretching back thousands of years. Analyses of Syria and Iraq, together with studies of the violence at Ayodhya and Preah Vihear in North India and at the Thai-Cambodian border, form part of a now well-established line of scholarly inquiry concerning the politics of heritage, one that seeks to explain why the past and its material legacies come to be contested, disputed, and, in some cases, violently fought over.[5]

In this book I want to take a different path to this topic, not looking at instances of contestation and violence but rather considering cooperation and the role that history and heritage play therein. My contention is that this other half of the "politics of heritage" has been greatly neglected within a field of critical inquiry that has read the political in terms of enmity, hostility, violence, or hegemony and nationalism. Although the latter remain in view here, they are couched in an altogether-different reading of the political, one that seeks to unearth the complex processes and consequences that arise when history and cultural heritage are called upon to facilitate trade and diplomatic relations, open borders, build intercultural dialogue and, as we will see in the case of the Silk Roads, even help shift the geopolitical landscape of global transport and energy markets. This book is not an international relations or political science text, but it does draw on the broad understanding of these fields that international diplomacy, collaborations, alignments, and partnerships are inherently political and politicized. In so doing, it seeks to provide a more critical analysis to the complexities of culture and history in international affairs, themes that too often remain marginal or are analytically parked as "soft power."[6] As Brook, van Walt van Praag, and Boltjes note, history is intimately connected to international relations in Asia, and yet the insights of the discipline too often remain overlooked.[7] The focus here is not history but rather its uses, or, to be more specific, the use of a narrative—one that might be best described as an aesthetic, an evocation—for political and economic ends. *Geocultural Power* thus builds on Samuel Huntington's recognition that culture, religion, and history have all become integral to global politics, just not in the ways he envisaged.

My interest in this topic was triggered some twenty years ago, when conducting research on the postconflict reconstruction of Cambodia. Listed as a World Heritage Site in 1992, the temple complex of Angkor in the northwest of the country emerged as a totem for societal and cultural recovery after a generation of violence, whereby civil war and genocide arose out of the Cold War conflicts enveloping Southeast Asia from the 1950s onward.[8] As Cambodia opened up in the 1990s, organizations from more than twenty countries offered assistance in heritage conservation and archaeology at Angkor. In what was an uplifting arena of aid, goodwill, and extraordinary generosity in spirit and resources, I became aware that this flurry of international assistance from governments, private donors, universities, and intergovernmental and nongovernmental agencies was underpinned by a highly complex, and rarely spoken about, moral and political economy. Guilt partnered trauma, asymmetry defined partnerships, and strategic interests invariably accompanied the giving and receiving of aid. Indeed, the entanglements and power relations of gifting that Marcel Mauss so eloquently identified underpinned the language of cooperation and collaboration, as well as the partnerships between organizations and between countries in the reconstruction of Cambodia.[9] Disentangling such power dynamics, forms of appropriation, and instances of the devastation of war and extreme suffering being exploited or capitalized upon proved a sensitive and oblique line of inquiry. Nonetheless, identifying the appropriate vantage point for understanding the complex cultural politics at play in the heritage diplomacy and international cooperation of postconflict reconstruction provided a valuable platform for grappling with the Silk Roads of the twenty-first century.

The case of Cambodia, and the forms of aid it received, is indicative of how culture and heritage are now being used to maintain or build relations with others where histories and cultural pasts overlap. In Asia and the Middle East, distinct shifts have occurred in this area in the past decade or so, as Abu Dhabi, China, India, Qatar, Saudi Arabia, Singapore, South Korea, Thailand, and others have emerged as significant exporters of assistance and aid in archaeology, museums, as well as in the policies and technologies that constitute modern conservation practice.[10] Japan has long couched such cultural-sector projects in a language of peace as part of a low-key "omnidirectional" diplomacy designed to rebuild goodwill in the wake of World War II.[11] Indeed, as the international coordinating committee for Angkor was set up in the early 1990s, Japan took the seat as one of its cochairs, along with France.

As India and China compete for economic and political influence in their region, such cultural-sector collaborations enable some very specific narratives of nation and history to be advanced. International aid is commonly underpinned by discourses of civilization, whereby both countries use a language of shared heritage to reinforce the idea that their "great cultures" stretched much farther than the boundaries of the nation-state today.[12] Now firmly ensconced in the infrastructure planning of cities and tourism economies, as well as community based and postdisaster or postconflict reconstruction initiatives, culture and heritage have come to be deeply entangled in other forms of "hard" aid. They have also become increasingly prominent in those discourses—sustainability, development, human rights, resilience—that underpin congresses and conferences around the world today and that shape the priorities of international cooperation funding structures. In pursuing such lines of inquiry, this book focuses on a project of extraordinary scale and ambition, one that promises to transform Eurasia in profound ways over the coming years and have an impact on ideas about culture, history, and heritage as mechanisms of diplomacy and cooperation more broadly.

Belt and Road

For some time China's political scientists and economists have been engaged in a debate on the challenge of maintaining economic growth and political stability over the longer term. In 2012 Lin Yifu and Wang Jisi, two Peking University scholars, argued that an economic model based on exports and development led by domestic investment was running out of steam.[13] They argued that China should invest the growing domestic surplus in emerging economies and use the country's export capital to develop those markets. Reeves suggests that this argument formed the genesis of a strategic proposal presented by the Chinese president Xi Jinping during a speech delivered to the government of Kazakhstan in September 2013.[14] The speech is now widely regarded as the launch of the Belt and Road Initiative (BRI). A month later Xi traveled to Jakarta, delivering a second speech on regional trade and investment, this time to the Indonesian government. But as Reeves points out, in their initial form, Central Asia and Southeast Asia remained separate, subregional initiatives. By the end of 2013, however, senior members of the Chinese leadership began to view them as a single strategy, whereby the two halves found integration through an idea of historical connectivity.[15] In the rapid evolution of Belt and Road, the development of economic ties to Central Asia and beyond constituted the "revival" of the land-based trade routes between East and West that had been established through the ancient Silk Road. In parallel, twenty-first-century engagement with Southeast Asia marked the reconstitution of the maritime trade networks of bygone centuries. China, it was proclaimed, would work with its partners to revive a Maritime Silk Road that extends down through Southeast Asia and across the Indian Ocean to East Africa.

Yí dài yí lù, or One Belt One Road (OBOR), thus became a paradigm of trade and investment stretching across more than sixty countries and incorporating almost two-thirds of the world's population. The project was also conceived to address a number of pressing domestic agendas, namely energy security, the development of periphery areas in China's northwest, as well as the economic reforms required for this to happen, an issue that was tabled during the third plenary session of the Eighteenth Chinese Communist Party Central Committee in late 2013. Belt and Road involves building new trade zones in Xinjiang and harnessing outward investment strategies across Central Asia.[16] Over the course of 2014, these domestic and foreign agendas were further integrated, such that the ethos and political ideologies of the national rejuvenation project, Chinese Dream, were extended into the realm of foreign relations. In 2015, Zhang Gaoli, at that time vice premier, chaired a newly formed high-level working group of five senior members of the party leadership. The composition of the team and its close ties to Xi were a clear indication of the importance and ambitious scope of the initiative.[17] Subsequent speeches led to an expansion of the themes brought under the umbrella of Belt and Road, a situation amplified by the use of the historical transcontinental Silk Road as a metaphor for regional trade networks, cultures of diplomacy, and the harmonious relations that come from the free movement of ideas, people, technologies, and goods (fig. 1.1). The Belt and Road Initiative, or BRI, as it has been subsequently renamed, thus places China at the political, economic, and cultural center of a vast geography of overland and maritime connectivity, leading analysts such as Wu Jianmin to describe it as "the most significant and far-reaching initiative that China has ever put forward."[18]

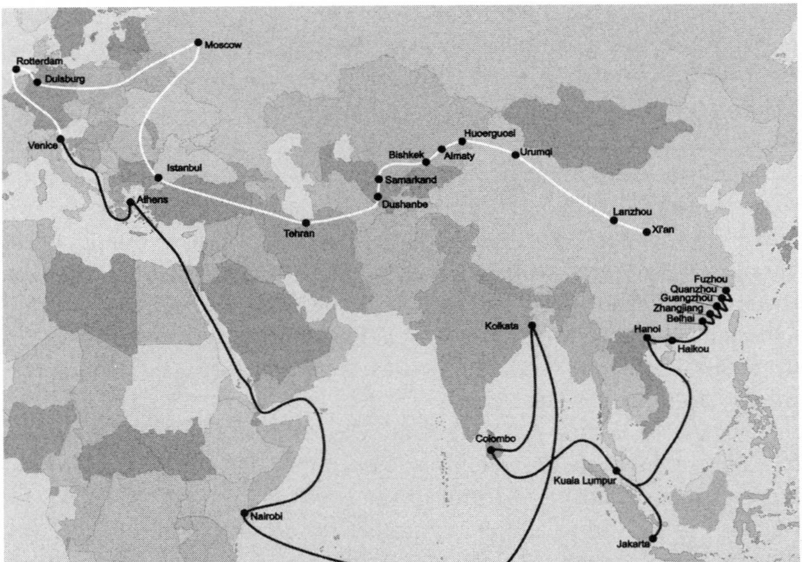

1.1 A typical representation of the Silk Road Economic Road and 21st Century Maritime Silk Route Economic Belt. Courtesy of Toyah Horman.

In March 2015, the Chinese government's Ministry of Foreign Affairs and Ministry of Commerce laid out the five-point Visions and Action Statement of connectivity and cooperation for "policy coordination," "facilities connectivity," "unimpeded trade," "financial integration," and "people-to-people bonds."[19] The 21st Century Maritime Silk Route Economic Belt and the Silk Road Economic Road were thus conceived to build corridors of development stretching across national boundaries via networks of rail, road, airports, pipelines, container shipping ports, as well as new, strategically located cities and special economic zones. The Visions and Actions Statement also proposed transboundary collaborations across multiple sectors, spanning energy, telecommunications, education, law, medicine, transportation, and culture. The Asian Investment Infrastructure Bank (AIIB), which came into operation in early 2016 with headquarters in Beijing and an initial operating budget of $100 billion, will help finance such initiatives. Moreover, $40 billion was set aside for the Silk Road Fund, with a further $124 billion committed in May 2017 as part of the first Belt and Road summit held in Beijing.[20] Fifty-seven countries committed their signature to the AIIB in 2015, with a further thirteen signing on two years later. In its initial conception, China held 26 percent of the bank's voting rights. Over the longer term, significant funding is also expected to come from the China Development Bank, which in 2015 declared its commitment to Belt and Road countries would be in the region of $900 billion.[21] It is widely agreed that Beijing is looking to maintain long-term growth in the Chinese economy through international trade and capital investments. In mid-2018 Michele Ruta, of the World Bank, cited changes in global value chains since the 1990s to argue that Belt and Road represented the next phase in China's ever-growing impact on worldwide trade. He anticipated it would continue to lower impediments to trade, lead to domestic policy reforms among BRI countries, and yield significant infrastructure investment dividends.[22]

It has also been suggested that Belt and Road advances a strategy to internally redistribute the wealth generated through urbanization by opening up the inland regions of China's northwest. The stability of Xinjiang and Tibet remains of paramount importance to the Communist Party, and transforming the cities of Urumqi and Kashgar into commerce and transport hubs represents a continuation of efforts to integrate minority groups, including Muslim Uyghur communities, with the rest of the nation. The government has long attributed instability in Xinjiang to economic factors rather than questions of political and cultural sovereignty. Attempts to address underdevelopment thus stretch back decades. Most notably, the Great Western Development Initiative, launched in 1999, focused on the economic and infrastructure development of six provinces, five autonomous regions, and the municipality of Chongqing.[23] This included a number of urban, agriculture, and communication projects across Xinjiang in an effort to rectify regional disparities in living standards. Interestingly, as far back as 1994, Premier Li Peng proposed establishing a "New Silk Road" for such purposes, connecting China up with the Middle East.[24] The further expansion of such policies across Xinjiang through Belt and Road are explored in chapter 4 via a discussion of Silk Road World Heritage nominations as corridors of conservation and development.

For Beijing and the governments of Central Asia, the fear of Islamic fundamentalism entering via long—and in many cases porous—frontiers has led to a form of state governance that integrates remote and border cities into the body politic of the nation-state through large-scale construction. In this regard, BRI represents a continuation of policies established under the umbrella of the Shanghai Cooperation Organisation (SCO), which since 2003 has targeted the three key priority areas of terrorism, extremism, and separatism. With governments committed to the idea that connectivity reduces suspicion and promotes common prosperity, a political mantra of peaceful trade and religious, cultural, and economic exchange directly addresses their concerns about civil unrest, both within and across national boundaries. The 2010 riots in Osh, Kyrgyzstan's second-largest city, which pitted an Uzbek minority against the Kyrgyz majority, illustrated how tensions between ethnic groups in the Fergana Valley can spill into violence, as well as the limited capacity states have to respond. In November 2015, Nursultan Nazarbayev, then-president of the Republic of Kazakhstan, chose UNESCO's Paris headquarters to announce the country's new Academy of Peace, stating, "We can best counter extremism through inter-cultural and inter-religious dialogue."[25] In June 2018, in the same week that the Group of Seven met in Canada, the seven members of the SCO also gathered in Qingdao, China. There they signed a Belt and Road Initiative joint declaration on cooperation in the areas of security and trade. Narendra Modi took on the Donald Trump–like role as the outlier of the group, refusing to sign the agreement on behalf of India.[26]

Analyses of Belt and Road have also honed in on Beijing's desire to secure its long-term energy security interests, with pipeline, port, and rail corridors through Northwest China and Yunnan seen as pillars of a strategic plan for scaling up and securing gas and oil supplies from the Middle East and Central Asia. The hard infrastructure elements of BRI

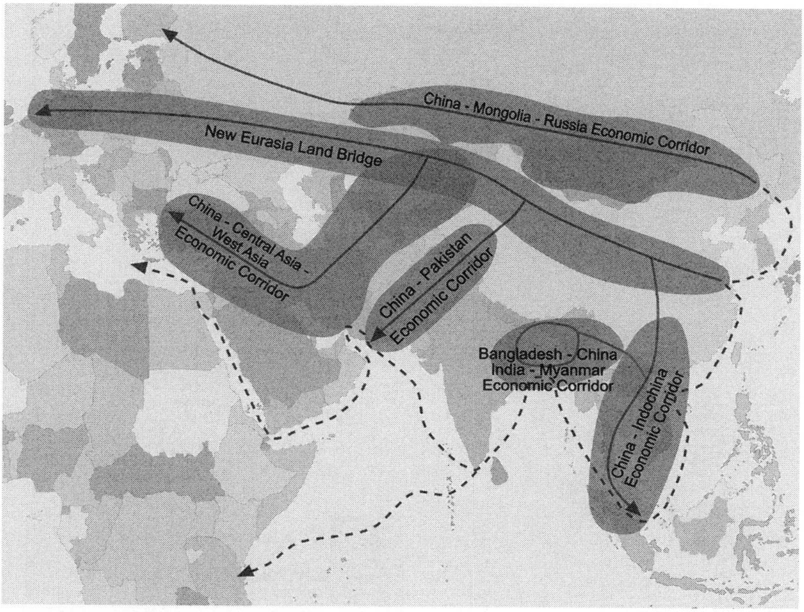

1.2 Belt and Road economic corridors and Eurasian land bridge. Courtesy of Toyah Horman.

are thus distributed across five economic corridors and an overarching Eurasian Land Bridge (fig. 1.2). Five years after launch, it is clear that major developments are being made along each of the corridors. To cite just a few examples, as part of the Bangladesh-China-India-Myanmar Corridor, the Chinese government has signed multibillion-dollar investment agreements for the development of deepwater ports in Chittagong, Bangladesh, and Kyaukphyu, Myanmar.[27] As Calder and Devonshire-Ellis note, both countries offer considerable strategic advantage to Beijing. Located in the Bay of Bengal, they provide China with an access point to the Indian Ocean, thus circumventing the costs, risks, and time involved in shipping oil and gas from the Persian Gulf via the Straits of Melaka and South China Sea (fig. 1.3).[28] Financial assistance for an $8.7 billion port and trade zone in Chittagong thus forms part of a much-larger strategic investment in Bangladesh centered on the construction of friendship bridges, airports, power stations, and a $3.1 billion rail corridor.[29]

To advance the China–Indochina Peninsula Economic Corridor, the 2015 Melaka Gateway Project featured a $10 billion investment loan from Beijing to develop the largest deepwater port in Southeast

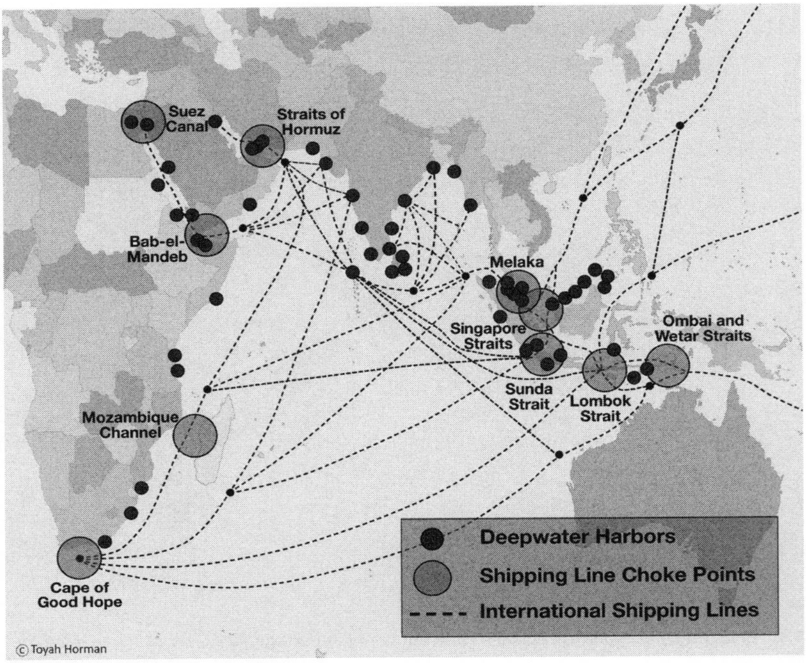

1.3 Security "choke points," deepwater port facilities, and international shipping lines of the Indian Ocean region. Courtesy of Toyah Horman.

Asia over the following decade.[30] In 2016 China continued its developmental aid to Cambodia, with Xi Jinping visiting Phnom Penh to sign Belt and Road agreements for the construction of road and energy infrastructure and special economic zones worth hundreds of millions of dollars.[31] For the China–Central Asia–West Asia Economic Corridor, Iran has already emerged as a major partner. Bilateral trade with China increased from $4 billion to $13 billion between 2003 and 2013, a trend that is expected to increase significantly as BRI projects and agreements come into operation.[32] Once again, massive urban development, energy, and transport infrastructure projects form the basis of agreements tying Iran's economy to Chinese companies and investors.[33] With the lifting of international sanctions on Iran, the two governments agreed to increase bilateral trade in early 2016 to $600 billion within ten years.[34] In Pakistan, China has provided a loan of more than $1.5 billion to cover 85 percent of the cost of an oil pipeline that connects to its Iranian border.[35]

For many observers, such developments need to be interpreted as historically significant shifts in the great power landscape of the twenty-first century. The marker of a great power is its ability to determine the norms and ideals of the international order. For David Arase, Belt and Road represents a significant advance in China figuring out how to restructure the world around it, a process that involves weakening the structural advantages in world trade and finance held by the United States.[36] It also speaks to a desire to move the dial in the balance of power across Eurasia by displacing the influence of Russia in the Caucasus and Central Asia.[37] As we will see in the following chapters, Beijing has declared that China's economic rise, together with the Chinese Dream, can be shared by others in the region through "win-win" cooperation. Moscow is keen to expand the types of strategic partnerships it established with Beijing in the 1990s, with Vladimir Putin proposing a project that would see the two countries lead a regionwide program of pipeline, rail, and highway connectivity.[38] But a strategy that builds regional economic integration and advances the internationalization of the renminbi through trade and multilateral banking structures has led many to speculate about the broader political goals of BRI.[39] I return to the relative military strengths of the region shortly, but it is first worth considering the diplomatic dimensions of Belt and Road, as they are pertinent to the chapters that follow. Arase suggests that Beijing is "laying out a vision of regional order that fits its unique set of cultural norms, political values, and core interests."[40] This is undoubtedly a complex matrix of values imbued with particular readings of history, wherein China is looking to reclaim its great power status—both culturally and politically—and overcome the bitter legacy of the "century of humiliation," which ran from the First Opium War of 1840 through to the defeat of the Japanese in 1945. In that regard, the respect that China is seeking internationally bears heavily on the nature of its trade and diplomatic relations, as well as on its conduct in international organizations. Su-Yan Pan and Joe Tin-Yau Lo are among those who have considered this dynamic at length. Together with Anthony Miller, they argue that China's rise requires a reconceptualization of international power relations and ideas about soft power.[41]

In recent years, a number of commentators in the West have considered the "rise of China" as a great power transition, wherein history suggests that a future of conflict inevitably beckons, the so-called Thucydides trap.[42] Others, however, argue that such a thesis misses the significance of the historical rhythms of a peaceful Chinese political culture.[43] Pan and Lo, for example, take up the tributary tradition, a concept that remains controversial among historians and even more contentious when applied to affairs today.[44] With its origins in the Han dynasty (206 BCE–220 CE), the tributary system involved foreign states paying tribute to the Chinese emperor in accordance with prescribed Confucian rituals. This led to tributary states in the region, such as Java, Korea, Sri Lanka, and Vietnam, adopting various aspects of Chinese culture and conducting trade within a structure of peace predicated on the acceptance of the superiority of Chinese civilization. Those who subscribe to the thesis of Chinese expansionism as one of harmonious relations argue that the system also depended on China's portraying of itself as an instrument of benevolent governance and trade. Pan and Lo suggest such histories offer a productive framework for interpreting Beijing's approach to the South China Sea. As we will see at various points through this book, others take a more critical stance toward both contemporary events and the models of benign, even benevolent, pasts deployed to interpret them. At this point, however, it is important to note the role that culture and history play in China's ambitions to secure influence internationally.[45] Indeed, I would argue that as this history has been articulated as national heritage over recent decades, it has also been deployed internationally as a force of persuasion and attraction, the hallmarks of soft power. The political significance of China's cultural past has further solidified through Xi Jinping's ascendancy to the position of general secretary of the Chinese Communist Party (CCP). At the heart of the Chinese Dream, an idea he has used repeatedly, is the concept of rejuvenation and renewal. Crucially, and as noted earlier, it is a political discourse conceived to reach both domestic and international audiences, one that morphed into the Asia-Pacific Dream in 2014.[46] Beyond its economics components, the Chinese Dream represents a desire to revive the Middle Kingdom, whereby a pride for both the achievements of the past four decades and five thousand years of history once again place China at the center of the universe.[47] A number of observers have thus pointed to 2014–2015 as a period of significant transition, away from Deng Xiaoping's "keep a low profile" strategy to a foreign policy dictum of "striving for achievement" that is reinforced by more proactive forms of neighborhood diplomacy.[48] Such developments open up fascinating questions concerning how the notion of historical Silk Roads bear on this dyad of China's sense of its "place" in history and the conduct of its international affairs today, an issue I return to in greater detail in chapter 7.

It is in this context that the cultural dimensions of Belt and Road have taken shape. The notion of reestablishing the maritime and overland Silk Roads locates China's economic rise within a much-deeper historical narrative, and thus internationalizes compelling ideas about the regional, even global, significance of Chinese civilization. Chapter 3 launches into this story by highlighting the use of Admiral Zheng He in contemporary diplomatic discourse. Zheng He is frequently presented in speeches as a diplomat who exchanged gifts and ideas, and thus as evidence of a long history of peaceful maritime engagement.[49] Subsequent chapters also highlight the subtler, but perhaps more powerful, ways in which ceramics, shipwrecks, silks, and historical urban environments communicate the reach of Chinese culture and civilization across the centuries. The degree to which other countries across Asia and beyond are accepting or resistant to such overtones is a complex issue. Many of China's regional counterparts, small and large, are nervous about the shifting power balance of East Asia and Beijing's military and economic ambitions. There are real concerns in Malaysia, Indonesia, Sri Lanka, Thailand, and Kenya about the growing presence of overseas Chinese and foreign direct investments. The argument that China has become increasingly assertive in its political dealings with regional partners has also gained weight in recent years. Andrew Scobell is among those observing that, "while China is relatively peaceful, it has begun acting more belligerently and more threatening."[50] The themes explored in this book reveal how the Silk Roads represent a strategic attempt to soften and alter the nature of China's engagements. Going beyond the rhetoric of economic win-win cooperation, the book sets up the idea that histories have been shared and societies have long been connected and culturally entangled. The discourse of BRI is to rebuild these bridges between governments and between populations. It is within this complex political space that a cultural past of routes and connections is being dug up, carefully restored, and put on display.

This book thus argues that we are seeing the past being used as a mechanism of great power diplomacy. More specifically, it maps out the various ways the Silk Road now serves as a platform for China to exercise its geocultural advantage. Here, then, the geocultural is akin to the geopolitical. In invoking the *geo* prefix, I wish to signal how power is accumulated by organizing and operationalizing geographical space. And, given that processes of heritage are about ordering the past into particular spatialized narratives, we see how the "ancient" Silk Roads have become the apparatus that orders people and places—potentially separated by vast distances—into modes of cooperation across multiple sectors. The threads of analysis here reveal how geocultural power comes from being the author of world history and also from being the architect of the bridge between East and West. Although much of the book focuses on the international dimensions of Belt and Road, an examination of Xinjiang reveals how the Silk Road narrative also extends more familiar modes of heritage governmentality at the subnational level. It will become apparent at different moments in the book that as India, China, and others vie for influence and primacy across Asia, different aspects of culture, religion, and history are strategically mobilized to win friends and build loyalties, and to legitimize expansionist ambitions to public audiences, both at home and abroad. The analytical themes pursued here pull these together as a geoculture, one that is doing considerable political work at a time when international power is accumulated and exercised through cross-border trade and connective infrastructures. In presenting a biography of the Silk Road in the chapter that follows, I argue that China is far from alone in having appropriated the concept for political and strategic purposes. But what we see in Belt and Road is Asia's ascendant power framing its international interests in a story of regional, even global, connectivity, at the center of which lies Chinese civilization. In this regard, the Silk Roads offer China a new form of geocultural power. But it will become evident that a discourse of twenty-first-century trade and cooperation based on historical networks of silk trade, seafaring, market cities, and cross-cultural encounters has encouraged other governments to use their own cultural connections of previous centuries in their crafting of trade and diplomatic relations today. Iran, Turkey, Sri Lanka, and the Arab states of the Persian Gulf are among those looking to Belt and Road as a vehicle for not only securing international recognition for their culture and the historical importance of their civilizations but also for using that sense of history to create political and economic loyalty in a region characterized by unequal and competing powers.

The Dream of an Integrated Eurasia

Many of the priorities identified under Belt and Road represent a continuation of developments that have transformed Asia over recent decades. One of the most dramatic trends has been the degree of physical and economic integration of the region. In what has been described as "Asia's era of infrastructure," there has been an unprecedented growth in transborder connectivity over the past thirty years, with projections suggesting that around 60 percent of global investment in infrastructure will be made in Asia in the coming decade.[51] Since the end of the Cold War, various multilateral and unilateral attempts have been made to build a more integrated Eurasia. To cite a few examples, back in 1992 UNESCAP set out to upgrade rail and road networks for the entire region, with landmark moments coming with the adoption of intergovernmental agreements on highway and railway projects in 2003 and 2006, respectively.[52] Programs were launched for Central Asia, with the city of Baku hosting in 1998 the interestingly titled conference TRACECA—Restoration of the Historic Silk Route on regional rail development.[53] Indeed, as Nadège Rolland notes, by the late 1990s Central Asia had emerged from decades of isolation, with various initiatives beginning "to give a concrete shape to the idealized vision of an interconnected and economically successful Eurasian continent."[54] For countries like Japan and South Korea, the post–Cold War era brought new diplomatic challenges and a shift in their geographies of developmental aid toward Central Asia. In 2004 South Korea announced its own Iron Silk Road Project, with Japan continuing its program of regional official development assistance with the Arc of Freedom and Prosperity Initiative about two years later.[55] Farther south a number of Southeast Asian countries got together, with support and funding from the Asian Development Bank, to create the Greater Mekong Subregion, or GMS, in the 1990s. Primarily oriented around the development of regional economic corridors, the GMS also provided an important platform for developing cross-border transportation links.[56] As Rolland notes, though, funding for such projects from the European Union, Asian Development Bank, or World Bank has often met resistance by governments across Asia suspicious of their pro-democracy agendas.[57] Indeed, the gradual emergence of a vast network of infrastructure projects—all of which have dramatically increased the flow of capital, goods, people, ideas, and technologies—has played a significant role in the story of major political change since the end of the Cold War. In the past decade or so we have also seen significant growth in intraregional aid, most notably from China, which has supplemented and in some cases replaced existing structures of bilateral and multilateral aid from the West. As Emma Mawdsley has argued, such trends are set to continue, with the AIIB injecting billions of dollars of investment funding over the coming years.[58] The twenty-first century, it is suggested, will be the "Asian century," driven in large part by the rise of China and India. Looking across the Asian continent, however, economic development has been extremely uneven in recent decades, with countries experiencing sustained periods of double-digit growth neighboring those that are weighed down by major governance, demographic, health, education, and/or infrastructure challenges. Belt and

Road has thus been thrust upon a region experiencing increasing inequality and asymmetry, both within and across borders.

The political geographies Belt and Road enters into also continue to change in response to the invasion of Iraq by the United States in 2003. In the Middle East, leaders have been toppled, states have failed, and new states have come into existence. Syria has been engulfed by extreme levels of violence in a civil war that, at the time of writing, has lasted more than eight years. In East and South Asia there is a rapid upturn in state militarization. China's growing economic and military strength has become a cause of considerable concern for Japan, Taiwan, India, and others in the region. The vast majority of the analysis dedicated to this issue has focused on East Asia and the South China Sea. More recently, however, attention has turned to the Indian Ocean region as a space of geopolitical rivalry and competing security interests. For some years, analysts in India and the United States have been scrutinizing China's growing presence in the region, with questions raised concerning the construction of deepwater ports along the coastlines of various Indian Ocean countries. Debate has surrounded whether this so-called string of pearls of commercial ports constitutes a future infrastructure of naval bases encircling the Indian subcontinent. But to understand the Indian Ocean region today requires looking beyond the action of any one country and grasping the complex matrix of actors that regard the region's sea lanes, commercial shipping infrastructures, and submarine natural resources as critical to their energy and national security (fig. 1.3). The control enjoyed by the United States in the post–Cold War era has given way to a situation of great power rivalry as India and China have built up naval forces and created spheres of influence through partnerships with smaller states. A number of the bilateral relations reflect key points of vulnerability, often referred to as "choke points" in the interregional flow of Persian Gulf oil and commercial shipping. For India, the markets of East Asia offer critical trade opportunity. As Pattanaik notes, more than three-quarters of India's trade, and more than 90 percent by volume, is maritime based.[59] This, together with Beijing's strategic investments in South Asian and East African countries, has had the effect of transforming India's foreign policies. Prime Minister Modi's Act East strategy steps up New Delhi's desire to become an Indo-Pacific power with influence over the security and trade dynamics of the region.[60] India is therefore among those countries that are reorienting and increasing military spending around a strategic maritime future.[61] The Indian Navy has responded to China's presence in Pakistan, Sri Lanka, and Bangladesh by leading joint exercises with regional counterparts, with the aim of becoming the region's preeminent security provider. Formulated as Security and Growth for All in the Region, or SAGAR, in 2015, this new foreign policy discourse emphasized "common goals," security as a "collective responsibility," and codevelopment of "integrated measures" for the shared "maritime home" of Indian Ocean nations.[62] New Delhi's investment in the Indian Navy as a mediator of regional diplomatic and security relations represents a major shift from its Cold War isolationist stance and from the mind-set that the dangers of militarized conflict would be land based and emanate from the north.

In gathering such issues together, Bouchard and Crumplin argue that the "Indian Ocean will be at the forefront of world geopolitics and global strategy" for decades to come.[63] And as Robert Kaplan suggested back in 2009, the region not only encapsulates all the dimensions of the great-power rivalries—China, India, and United States—but also brings them together in a physical landscape that will continue to have legal and governance frameworks wholly inadequate for accommodating the competing interests of the many countries deeply invested in the region's future. Pertinent to the themes discussed in this book is China's limited ability to project power across the region as a result of the geographies of the Indian Ocean, most notably extensive distances between China's bases and their impediments to fortification.[64] China's military deficit to the United States and India, with the latter enjoying significant geographical benefits, means that Beijing is mostly likely to continue to secure influence through economic investment and different modes of diplomacy. Its strategic imperative centers on maintaining its sea lines of communication (SLOCs), and the connections between East Asia, the Persian Gulf, and through to the Mediterranean.[65] The overland connections across Pakistan, Myanmar, and Bangladesh play a crucial role here, with plans for port facilities and refineries in Gwadar, Kyaukphyu, and Chittagong, respectively, enabling ships to connect with the pipeline and rail infrastructures now under the umbrella of the overland Silk Road. China's plans for expanding its footprint in the Indian Ocean also build on existing infrastructures such as the Yunnan-Yangon Irrawaddy road, rail, river corridor, which has facilitated cross-border trade between southern China and Myanmar for more than a decade.[66] Similar overland-maritime connections in Pakistan, Iran, and Bangladesh all mean that it is misleading to treat the two elements of Belt and Road as separate spheres of connectivity and trade.[67]

The installation of the Trump administration in early 2017 heightened tensions with China in the South China Sea, as standoffs and provocative gestures of strength from both sides increased. Ongoing missile tests by North Korea, followed by a Singapore summit and trade tariff war in 2018, further point to uneasy futures. Against this backdrop, Asia's governance landscape remains characterized by a paucity of intergovernmental bodies and pan-regional mechanisms capable of mediating state-based disputes.[68] Together, then, these various changes have profoundly altered the ways in which the past and the discourses of history and heritage are mobilized and used across the region. Collective ideas of memory and heritage act as mediators of cooperation and hostility in an international relations environment that is significantly more complex than it was twenty years ago, and their presence in this arena continues to change quickly. In East Asia, museums, memorials, and tourist sites have become weapons of public diplomacy and propaganda, raising the temperature on the ways in which World War II is remembered and commemorated.[69] And as we saw in the discussion of Islamic State, the widespread destruction of heritage formed part of the IS's foreign policy strategies. The themes and lines of critical inquiry pursued here, however, revolve around international cooperation and the fascinating role the past and its material legacies are taking on in this ongoing, and seemingly accelerating, path to Eurasian connectivity.

Heritage Diplomacy

To interpret trends and connections, the book develops the concept of heritage diplomacy. As outlined elsewhere, heritage diplomacy seeks to understand how cultural pasts and material culture become the subject of exchanges, collaborations, and forms of cooperation within wider configurations of international relations, trade, and geopolitics.[70] Today, hundreds of millions of dollars are spent every year on preservation, with governmental and nongovernmental institutions, universities, and a multitude of professional agencies clustered in networks of cooperation. To contextualize this, heritage diplomacy reveals the various ways in which cultural forms inherited from the past and discourses of preservation are both constituted by and contribute to the interplay of cultural nationalisms, international relations, and institutions involved in the practices of global governance. Critical theory on UNESCO's activities today or on the domestic policies of states and their institutions often casts activities and worldviews in a social and political vacuum. It is also the case that many of those involved in the actual practice or policy of heritage preservation regard their work as apolitical. Closer inspection reveals this is rarely the case, particularly when international collaborations are involved and where funding crosses borders. My sense, then, is that considerable work still needs to be done to understand the multilayered, cross-sector political and economic forces that shape preservation and how institutions involved in cultural and heritage governance operate.

Today the state appears to be an enduringly powerful actor. Thomas Weiss has argued that international bodies such as the United Nations, including UNESCO, remain "flimsy" in shaping international affairs, and this is particularly the case in Asia, where the state continues to be a key force exerting its will on the cultural past.[71] The incorporation of cultural-sector cooperation into other streams of humanitarian and developmental aid, as noted earlier, and trade and diplomatic relations means that questions surrounding neo-imperialism remain as pertinent as ever. To begin addressing such issues, the concept of diplomacy is most productive when considered in broad and critical terms. Cooper, Heine, and Thakur are among those suggesting we need to pursue far more analytically expansive readings of diplomacy, analyses that look far beyond government departments, embassies, and ambassadors to include the conduct of nongovernmental and philanthropic agencies, the corporate sector, state-funded bodies, and intergovernmental organizations.[72] To account for how this area has grown in scope and complexity in recent years, Murray and his coauthors propose that diplomacy is now thought of as the "institutions and processes by which states, and increasingly others, represent themselves and their interests to one another in international and world societies."[73] This book foregrounds history and cultural heritage in such lines of inquiry, and in so doing, it addresses the Silk Road as a strategic narrative in the ways outlined by Miskimmon, O'Loughlin, and Roselle, and also by Wastnidge in his examination of Iranian diplomatic policies under President Hassan Rouhani.[74] Heritage diplomacy also challenges and, I would suggest, departs from conventional understandings of cultural diplomacy as the export of social and cultural goods for soft power. As a framework it seeks to understand how the cultural past sits in that interface between international relations and governance, involving multiple state and nonstate actors as it straddles sectors as diverse as architectural conservation, development, urban and infrastructure planning, and international trade. By recognizing heritage as an instrument of spatial and social governance in this way, we can trace the different ways it comes to be exerted as a form of hard power. But here the book also focuses on the contemporary international politics surrounding objects of antiquity. With much of the focus in this area falling on disputes over ownership, with tensions between Britain and Greece over the Elgin Marbles among the notable examples, the chapters that follow reveal a quite-different politics at play when the discourse of history and heritage moves to ideas of shared and entangled pasts.

Recent thinking about networks in the social sciences informs this analysis, in the sense of both understanding globalization as the conglomeration of various networked hubs and flows as offered by Castells, Sassen, and others, as well as ideas about assemblages and human, nonhuman relations as identified in actor-network theory.[75] Together these offer insights into the spatial dimensions of power and politics, as well as how social and material worlds come to be mutually ordered in certain ways at particular moments and the reasons such orderings might remain resilient or fall apart. Belt and Road represents such an ordering on a continental scale, one that is based on an interpretation of another spatially and discursively vague ordering of history, that of the "ancient" Silk Roads. Actor-network theory pluralizes who and what we think of as actors in the social world and the associations they create.[76] It is an ontology and methodology that places particular focus on the agency of nonhuman actors, and it takes the entanglement of the social and material as its starting point of investigation.[77] Applying such ideas to the field of archaeology, Ian Hodder has argued that entanglement leads to chains of dependencies between human and things, and things and things, whereby the "affordances" created between elements builds stability into the ordering of everyday life. In pursuing such ideas on a much grander scale, *Geocultural Power* reveals how histories and cultural artifacts from the past are being excavated and symbolically coded in ways that help ensure cities and entire countries are tied into the new networks of trade being established under Belt and Road. In other words, ceramics, shipwrecks, buildings, and even tiny fragments of silk cloth help afford participation in an emergent ordering of transcontinental trade and diplomatic relations. The new Silk Roads, then, are not just the bundling up of geopolitical power and geography but also the bundling of material objects from the past into a grand narrative of connectivity, past and present. Crucially, and as we will see, this happens in and along certain places and corridors. In this regard, the book brings cultural heritage into conversation with energy pipelines and roads, analyzed by Andrew Barry and Agneishka Joniak-Luthi, respectively, as unstable material actors in the creation of transboundary technical and political ties across Central Asia and Northwest China.[78]

The aim is not to unpack these assemblages of heritage diplomacy into their constituent parts, but to see how elements come to be entangled and are in flux in ways that help transform a vague and ambiguous narrative of the Silk Road into a physical ordering of relationships across multiple sectors. In other words, Buddhist caves, mosques, museums, and shipwrecks, together with airports, roads, hotels, pipelines, and deepwater ports, are among the heterogeneous elements being entangled through aspirations to order Eurasia's physical, economic, and political landscapes in extraordinary and historically unprecedented ways. The discussion of Xinjiang in chapter 4 suggests that, in certain contexts, this amplifies existing governmental rationalities. But in Deleuze and Guattari's assertion that assemblages are essentially productive, we are also reminded that as their constituent parts come together, new realities and new territorial organizations can emerge.[79] As we will see, the case of the twenty-first-century Silk Roads resonates with their concerns for deterritorialization and reterritorialization, whereby space, culture, and history are undone and remade in ways that embed new forms of power.

Geocultural Power thus explores the strategy of couching trade and political relations, energy and political security, in an evocative topography of history. The Silk Road has become one of the most romanticized and mythologized landscapes of modern times. In the West it evokes all the mysteries and fantasies of an exotic Orient, images of great civilizations that stretch across expanses of time and land, and of rich, colorful cultures shaped by the environmental conditions of mountains, deserts, and vast open plains. The romantic appeal of the Silk Road means that it remains an ever-popular framing for adventure tourism, restaurants, and cookbooks, as well as an array of musical, dance, and theatrical performances. In the world of Silk Road dining, restaurants in Moscow, Los Angeles, London, and elsewhere take one of two approaches to designating the origins of their food. In some cases, Afghan or Iranian dishes are elevated far beyond their national contexts; elsewhere, menus offer a range of fusion dishes using the vast geographies of Eurasia to playfully blend different ingredients and culinary traditions. For tour operators, a Silk Road imaginary enables them to market countries and cities in Central Asia—that are either unknown or perceived to be high risk by clients—as exotic, romantic adventures. Travel maps of the Silk Road sold on Amazon unfold to reveal a smooth cartography of northern and southerly routes, with lines highlighted by camel icons stretching across mountains and deserts, linking cities thousands of miles apart.[80] And the travelogues of Ibn Battuta (1304–1369) and Marco Polo (1254–1324) remain compelling stories of Silk Road adventure and discovery. Their accounts talk of great cities and of cosmopolitan communities. Chapter 2 gives a detailed account of how the idea of the Silk Road has evolved in the 140 years or so since Ferdinand von Richthofen first used the term in 1877 to describe trade routes from China to Europe via Central Asia during the Han dynasty (206 BCE–220 CE). The discussion explores how the geographies and timelines of the Silk Road have expanded in the modern era and considers the factors that have led to it being associated with certain values and ideas. Today, documentary makers, travel writers, and historians present the Silk Road as a history spanning two millennia and encompassing much of the Eurasian landmass, Indian Ocean, and seas of East Asia. A narrative of cross-cultural exchange, vast and open landscapes of adventure, and the romance of ancient civilizations, the Silk Road emerged in the twentieth century as the quintessential story of premodern globalization, a more harmonious world predating the rivalries and violence of nationalism, imperialism, and industrialized war.[81] Chapter 2 begins by noting the lack of consensus concerning where the Silk Road is said to begin and end, when it peaked and declined, and whether it is best framed as a story of regional trade or religious and cultural transmission. Perhaps more significant, there are also strongly divergent opinions among scholars as to the analytical merits of the term. James Millward thus suggests we view it as a history of "exchanges of things and ideas, both intended and accidental, through trade, diplomacy, conquest, migration, and pilgrimage that intensified integration of the Afro-Eurasian continent from the Neolithic through modern times."[82]

Not surprising, however, the story of the Silk Road has often revolved around the production and circulation of silk itself.[83] With China emerging as the first center of silk production, luxurious, lightweight textiles became the ideal gifts and trade items for long-distance travelers and merchants. As direct and indirect contact was made with southern Europe, the Roman Empire's adoration of silk grew rapidly. For the Romans, silk became a symbol of luxury, military prowess, and erotic femininity, and much scholarship has been dedicated to tracing the trade routes and intermediaries that sprung up to meet demand.

Xinru Liu, for example, extensively details the profits Byzantine traders made on the back of a Roman fetish for purple-dyed silk. To supply this craze, the Byzantines developed a state monopoly over production and sales, using silk yarn imported from China via overland routes that wove their way through Persia or by sea routes from India.[84] Liu thus suggests that "in this militarily weak but materially rich state, the silk textiles were used to build and strengthen royal and ecclesiastical hierarchies and were a very useful tool in diplomacy."[85] From the sixth century onward, Persia enticed Chinese consumers with different styles and designs drawn from the manufacturing techniques developed for making woolen items. With exports flowing back and forth, trade continued to thrive through the centuries as silk became integral to ritual, beauty, and luxury of the Christian, Islamic, and Chinese worlds. As an ever-constant high-value commodity, silk also became an important driver of urban development, as guilds and trading markets formed in cities across the region. In the Islamic world, the wearing of luxuries carried the risk of not going to paradise. Over time, however, codes emerged for the amount and design of silk that was permissible in garments. During the Abbasid Caliphate (749–1258), a tolerance for more opulent lifestyles meant that guilds were established in cities such as Baghdad to supply textiles for clothing and interior decoration.[86] One landmark account in Silk Road studies, Luce Boulnois's 1963 *La route de la soie*, pursued the international circulation of silk through to the modern period, wherein she identified the important role of Lille in nineteenth-century France, as well as events in early twentieth-century East Asia, such as the destruction of Chinese mulberry-tree plantations by Japanese bombers and shifts in demand caused by the introduction of nylon in the postwar period.[87]

The portability of silk made it ideal for transporting across great distances via camel and horse, and in this respect, it was far from alone. Indeed, other scholars adopting the concept of the Silk Road have stressed that an emphasis on silk belies a larger history of cultural transmission and trade.[88] High-value, lightweight commodities were pivotal to the creation of trade patterns between different social groups and to the success of market towns that linked producers and consumers separated by great distances. Over the centuries, caravans of camels and horses carried spices, herbs, incenses, precious metals, wine, carpets, and tea.[89] The popularization of commodity histories in recent years has led to a flurry of studies focused on singular items—cotton, tea, cumin, or paper—as mediators in the relations between empires, religions, and cultures.[90] For others, however, the real significance of the Silk Road concept lies not in its account of regional trade but in its pointing to the transmission of religion, culture, and technology over great distances, and in its highlighting of histories of nomadic culture across the Eurasian Steppe.[91] Along its routes passed scientific, engineering, and medical knowledge.[92] Gunpowder and paper technologies were exported from China, and ideas about weaving, vine cultivation, and glazing migrated outward from the Middle East and Mediterranean. At different times, cities such as Samarkand and Baghdad emerged as key hubs of connectivity, with merchants traveling from Afghanistan, India, Pakistan, and Sri Lanka meeting caravans traveling east-west. Remote monasteries and pilgrimage sites also benefited from the exchange of ideas and religious systems brought by long-distance travelers. Richard Foltz uses Silk Road trade as his point of departure for understanding the diffusion and interaction of different religious cultures across Asia. In contrast to the more familiar idea of religious empires or worlds, Foltz traces Zoroastrianism, Manichaeism, Nestorian Christianity, Judaism, Islam, and Buddhism as histories of entanglement and networks of connection.[93] Such an analysis of religion reveals the importance of Central Asia in the story of premodern globalization, an issue picked up in chapter 6.

Clearly, then, silk trade is only a tiny component of the flows and connections across Eurasia over multiple centuries. I have dwelled on it, though, as it has become a metonym for a number of ideas that carry weight here. First, across cultures there is the allure of the luxury of silk, a fragile textile of irreproducible and exclusive beauty. In both its materiality and its symbolism, silk not only gives the road a uniquely romantic aura but also signals the technological prowess and deep appreciation of beauty and craft of Chinese culture. But through the appellation "Silk Road," silk also was the quintessential commodity of long-distance trade. It carries a story of market towns and intermediaries, of open borders, and of a Chinese entrepreneurship that creates wealth by reaching new markets. Silk, it seems, offered unparalleled rewards for those prepared to undergo the hardship of traveling across desert and mountain and to maintain peaceful relations with foreign cultures. And it is these imaginaries—of trade, of a Eurasian past of harmonious relations, and of a Chinese civilization with refined aesthetics—that are implicitly called on in the discourse of reviving the Silk Roads for the twenty-first century.

As we will see in the following chapter, the birth of Silk Road studies occurs in the remote landscapes of Central Asia in the late nineteenth century, with the discovery of manuscripts, paintings, and other cultural artifacts. The items recovered from the region opened the window to a much larger and complex world of cross-cultural currents—a picture that has continued to take shape over the past century or so as archaeologists, art historians, philologists, ethno-linguists, and historians have interpreted and deciphered an array of material culture. The story of the Silk Road has been inscribed through a distinct repertoire of objects, such that buildings and shipwrecks, carpets and ceramics, maps and manuscripts, are the embodiments of connection and exchange.[94] Research on nomadic cultures has shed light on the cultural histories of the Eurasian Steppe. But as Kuzmina notes, this occurred in a context of evolving human ecology theory, with Soviet scholars of the 1960s, 1970s, and 1980s studying patterns of human settlement in relation to vegetation, precipitation, and broader, long-term shifts in the ecosystem of the steppe.[95] For Valerie Hansen, however, we can really fully comprehend the scale and nature of cultural transmission and exchange along the Silk Road only through a detailed study of the various texts that have been recovered and preserved. In laying out a new interpretation of Silk Road history in 2017, Hansen draws on the writings of medieval Chinese monks and other religious manuscripts, as well as letters written by travelers through the centuries and the various documents of bureaucracy that have survived the passage of time.[96] Hansen's approach extends an important theme in Silk Road historiography—the genre of travel writing, both historical and contemporary. Although the accounts of famed travelers such as Marco Polo, Ibn Battuta, and the seventh-century Chinese monk Xuanzang continue to inspire the routes and texts of travel writers today, Susan Whitfield takes an alternative approach, highlighting the archetypal characters—pilgrim, soldier, merchant, courtesan, and horseman—to tell the story of "life along the Silk Road."[97] As the following chapter demonstrates, a "golden era" of exploration and research beginning in the late nineteenth century has also become an essential part of the narrative. Tales of fabulous archaeological discoveries and arduous adventures across the hostile terrains of mountain and desert infuse the past with romance and intrigue. Morgan and Walters, for example, focus solely on the exploits of Aurel Stein, of which we will hear considerably more, to tell the story of "a desert explorer, the Buddha's secret library, and the unearthing of the world's oldest printed book."[98] The eventual opening of Central Asia's remote regions from the 1980s onward also enabled adventure-seeking journalists and filmmakers to "retrace" the footsteps of previous travelers, whether they be from the 1930s or the 1300s.[99]

Through the contention of the Maritime Silk Road, Belt and Road brings into focus cultural histories of the sea. Eurasia's extensive sea-based trade networks stretch back millennia, and much like the fabled overland Silk Road, constructing a historiography around a singular transcontinental route is highly problematic. The field of maritime archaeology developed later and at a slower pace than its land-based counterpart, particularly in Asia. Christopher Beckwith and John Miksic have reflected on the implications of this, arguing that the paucity of evidence concerning interregional maritime trade routes has led to imbalances in how connectivities across the Eurasian continent stretching over one and a half millennia have been historicized.[100] As Miksic notes, the ancient maritime networks involving goods traveling back and forth between Europe and East Asia have been widely overshadowed by the camel caravans and market towns of Central Asia.[101] For this imbalance to be rectified, Miksic acknowledges the importance of an equivalent terminology of evocative, imaginary pasts, proposing the Silk Roads of the Sea as a concept that can capture an expansive history of maritime trade and communication. Here, the *s* in *Roads* is charged with considerable analytical ambition, as archaeologists remain in the early stages of understanding the complexities of seaborne connections among China, Southeast Asia, and across the Indian Ocean to Persia and the Arab world. An argument developed in subsequent chapters centers on how this paucity of knowledge has, in part, enabled the notion of a Maritime Silk Road to gain currency in recent decades, with a narrative of the past crafted around those patterns of mobility evidenced in documentary records, shipwrecks, and artifacts found in port cities.

Indeed, one question this book asks is how the discourse of a Maritime Silk Road advanced by Belt and Road will give greater visibility to the different elements of Eurasian maritime history. Chapter 5 considers this issue through the lens of ceramics and shipwrecks, with chapter 6 broadening the discussion to the ways in which the Maritime Silk Road of the twenty-first century shines a spotlight on the histories of the Indian Ocean, a region too often overlooked in accounts of world history. Scholarship that uses world systems theory to interpret the ancient communication and trade networks of the Indian Ocean is cited in order to consider the degree to which heritage discourses and historiographies activated through Belt and Road are creating new ways of seeing Eurasian and world history. The language of reviving the maritime connections of previous centuries, however simplistic and overly romanticized, has created a platform affording histories of Swahili exploration, Persian and Arab merchant cultures, and regional port cities new possibilities of research and visibility. As cities and governments signal their place in these interconnected cultural pasts, an emergent Indian Ocean heritage represents a potentially important point of departure from the nation-state-centric framings of history and heritage that dominated the twentieth century. But as we will also see, the flurry of publicity and international media attention given to Belt and Road since its launch has given weight to narratives of maritime and overland histories of trade that are, not surprisingly, China-centric. One figure in particular has dominated the narrative, Zheng He. Born in 1371 in Yunnan, Southwest China, into a Muslim family, Zheng He commanded seven fleets during the early Ming dynasty, traveling across South Asia, the Arabian Peninsula, and West Africa (fig. 1.4). Zheng He's voyages have come to be widely celebrated in China and by overseas Chinese communities as peaceful, diplomatic missions. Politicians, museum curators, and documentary makers regale their audiences with stories of gift exchange and trade with far-flung kingdoms, with Zheng He delivering foreign rulers to the imperial court to offer tribute. This smoothing over of his voyages as benign and free of violence has taken on new diplomatic resonance around the region in the era of Belt and Road. China's attempts to reinstate the overland and maritime Silk Roads for the twenty-first century builds on ideas of regional diplomacy, friendship, and corridors of trade through which pass people, capital, goods, and ideas. It is a selective reading of

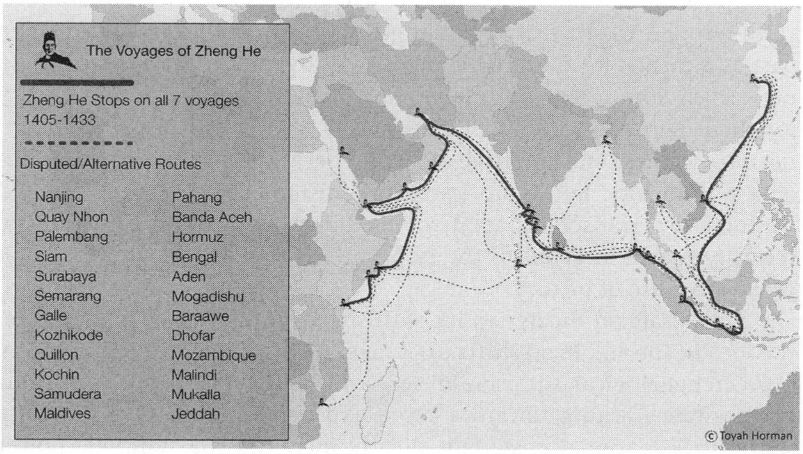

1.4 Zheng He's voyages and stops. Courtesy of Toyah Horman.

history designed to foster economic and political integration across the Eurasia region, one that Beijing is keen to promote in order to achieve a series of strategic goals.

With the Silk Roads invoked as a metaphor for connection, exchange, and integration across multiple sectors, Belt and Road has rapidly become a vast and extraordinarily complex arena of heritage diplomacy. The language and artifacts of cultural heritage are being deployed to smooth out the complexities of foreign relations and subnational politics, and to help build future diplomatic and trade ties. Enduringly romantic, the Silk Roads are being reimagined for the twenty-first century to smooth out history itself, erasing memories of conflict, great power rivalries, and past enmities. It is not uncommon for history to be glossed in this way to maintain cordial political and trade relations in the present. But what we see in the twenty-first-century Silk Roads is the explicit invoking of the past for such purposes, where the very spatial and temporal ambiguities of the overland and maritime Silk Roads are played with for expedient purposes. Museums, archaeological sites, fortified cities, shipwrecks, and the artifacts of trade and exchange from centuries ago are thus called on to smooth out labyrinthine, fragmented histories and events, pulling them together into a singular, integrative story of a shared past, a shared heritage. In a diplomatic dance of forgetting, episodes of violence, invasion, and bloodshed are left behind for a language of history and heritage that crosses borders in ways that directly align with the foreign policy and trade ambitions of governments today. In other words, heritage as a concept is being reimagined to enable countries and cities to strategically respond to shifting regional geopolitical conditions and use the past as a means for building competitive advantage in an increasingly networked Sinocentric economy. As countries appropriate this history for their own ends, they will continue to find points of cultural connection through the language of shared heritage to gain regional influence and loyalty. But as stories of long-distance trade are represented through two transcontinental routes drawn as lines on a map, little scrutiny has been given to the implications this may hold. This book probes this use of history.

To elaborate on the dynamics of the heritage diplomacy in play here, the idea of the Silk Road shifts discourses of governance and patrimony over archaeological sites, buildings, and landscapes from a previous nation-state framing toward a more expansive notion of a mutually shared transnational past. This greatly increases the symbolic power that material culture holds in diplomatic relations. The notion of a mutual, conjoined history moves cultural heritage away from being merely incorporated into existing or new diplomatic relations between governments and countries and toward becoming a mechanism for and of diplomacy in itself. A core theme of the book is understanding the political and economic forces that shape how the "history" of the Silk Roads comes to be revived. If it is not about restoring the transport infrastructures of camel caravans and Arabian dhows, then what is it about? Clearly, one of the key vehicles through which the Silk Roads are to be "revived" is the language of heritage. But as this is made to serve the goals of Belt and Road, the past is contrived, framed, and narrated in very particular ways. New places and material histories are brought into focus, the narratives through which they are interpreted are reworked, and the values by which they are given significance to contemporary society are redefined. In exploring such themes, my particular interest is how such processes work to build trade and political relations between countries, and how heritage is used to perform a second form of smoothing, that of diplomatic relations in highly uneven and fast-changing parts of the world. At this point, it is worth noting that the analysis here does not extend to philosophical or religious traditions, domains that, in varying degrees of abstraction, can indeed be understood as part of a country's heritage. Primarily, I seek to understand how objects, places, cultural performances, and discourses about past events and people, are being mobilized as part of the wider diplomatic relations and cooperation structures of Belt and Road. The final chapter of this book reflects on these developments at a more conceptual level by contrasting processes of smoothing to Anna Tsing's notion of friction. To get to that point, however, the five chapters that follow consider the heritage diplomacy of the new Silk Roads in different ways.

The chapter that follows presents a biography of the Silk Road concept. Starting with the discoveries made in Central Asia in the late nineteenth century, the chapter traces the development of a Silk Road discourse during the twentieth century. The discussion of its historiography is extended but understood within the politics of nation building, empires, and international conflicts of the twentieth century. The Silk Road is thus considered as an emergent geocultural imaginary, one that has been imbued with particular values and ideals in different parts of the world. The history of the Silk Road concept is deeply enmeshed in the geopolitics of territory and transboundary infrastructures, and its recognition internationally has in large part been contingent on the prevailing trends of modern historiography and heritage making—notably, the privileging of the nation-state. The transformations brought about by the end of the Cold War and economic reforms in China initiated by Deng Xiaoping are among the factors enabling a Silk Roads discourse to flourish and proliferate. Crucially, this biographical approach reveals the different ingredients that are being brought together in Belt and Road, and in so doing, it offers a more critical reading of what is at stake in the "revival" of the Silk Roads for the twenty-first century.

From there, chapter 3 begins to explore instances where the historical Silk Roads have been appropriated in the discourse of governments and politicians. In dissecting this language of revival, the chapter identifies the metaphors and ideas—harmony, friendship, trust, cooperation, and spirit—a combination of which have been adopted in recent times to nurture international relationships. It is argued that Belt and Road builds on and greatly extends certain historical narratives advanced by Beijing since the late 1990s. But as we will see, the Silk Roads have rapidly taken hold in the political discourses of leaders and diplomats across Asia and beyond as they seek to link up with emergent futures. Chapter 4 turns to the connections Belt and Road has fostered among infrastructure investments, the planning of transboundary development corridors, trade deals, and the conservation of World Heritage Sites, museums, and historical cities. The aim is to excavate distinct points of convergence, geographical and financial, and across different sectors rather than argue for lines of causality. The chapter discusses various projects and multibillion-dollar deals trumpeted as Belt and Road initiatives. In reviewing these, it is important to recognize that signed contracts do not necessarily lead to finalized projects. Belt and Road covers harsh and volatile landscapes, and agreements and construction sites will inevitably be abandoned as political and economic conditions change. Although it is impossible to accurately anticipate such futures, the chapter explores the template of cooperation and regional integration that has been put in place and critically discusses some of the broader developments Belt and Road has set in train.

The story of the Silk Roads is one of transport and the selling of items carried over great distances. Silk, spices, gunpowder, gems, ivory, metalware, furs, navigation technologies, and a plethora of ceramics were all carried across land and sea over the centuries. Those items that have survived help constitute stories of mobility, cultural exchange, and the cross-fertilization of ideas, and thus they have become key actors in Silk Road heritage discourses today. Chapter 5 examines this in relation to objects of itinerancy and the diplomatic affordances they offer. Particular attention is given to ceramics to illustrate how Belt and

Road imbues objects with particular values, even diplomatic qualities, as they are salvaged, displayed, and moved between exhibitions. The example of the Belitung shipwreck, its controversial salvage operation, and the collaboration between Oman and Singapore to reproduce the original ship are signposts to future trends in maritime heritage diplomacy. The chapter concludes by anticipating how a growth in interest in the Silk Roads in China and the region will affect the global trade in trafficked antiquities. It is suggested that in Belt and Road's construction of roads and rail and in its opening of borders an infrastructure for cross-border smuggling is being built, one that can more easily supply a demand for antiquities in East Asia that is likely to grow at an exponential rate over the coming years and decades.

Chapter 6 steps back to consider the ways a renewed interest in the heritage of the Silk Roads raises important questions about world history and how the identities of nations and cities are narrated. It is argued that Belt and Road represents an elastic and expansive international political economy that shines a light on histories that have yet to receive the attention they warrant. Recent research on Central Asia and the Indian Ocean are cited, as well as arguments concerning the importance of these regions to our understanding of world history. The chapter highlights the potential that exists for new stories to be told through academia, museums, and historical sites, and through the media. It also considers the degree to which the historical significance of Eurasia's cities, coastal and inland, is being remapped through the discourses of heritage advanced by Belt and Road. Finally, to provide some institutional context to such issues, the chapter also reflects on how forms of scholarly knowledge that feed into public discourses on Silk Road history are being rearranged and reassembled in an era of Belt and Road. It is argued that this shapes which issues and places are focused on and how they are interpreted.

The book concludes with three thematic discussions. The first returns to the analysis of the Silk Road as a geocultural imaginary and a form of geocultural power. It is argued that the Silk Roads provide China with a unique platform for exercising its geocultural advantage, and through history, culture, and heritage, we are seeing the forces of great-power diplomacy being exerted. An analysis that juxtaposes such processes to infrastructure projects, new trade deals, and so forth, reveals how the geoculture of the "ancient" Silk Roads intersects with the geopolitics of Belt and Road. This is followed by a broader discussion of the concept of smoothing and the complex ways this occurs through the heritage diplomacy of Belt and Road. In the final section the chapter reflects on some of the key factors that bear on the unfolding futures of Silk Road histories and the entanglements of trade, diplomacy, geopolitics, and cultural heritage. In both Belt and Road and the ancient Silk Roads we see two spatial arcs, each constituted through indeterminate networks of connectivity. These arcs are defined by ambiguities and possibility, frontiers and creativity. It is such factors that will determine where, when, and in what form the Silk Roads are celebrated and commemorated in the future.

Together, then, these six chapters explicitly move between tenses: past, present, and future. Despite the ongoing commitments made by Beijing and governments around the region, the precise pathways BRI will follow inevitably remain uncertain. The analysis offered here ventures into those uncertain futures, but it does so by observing some distinct trends and directions set in motion since 2013 and the years leading up to the launch of Belt and Road.

TWO

The Silk Road: An Abridged Biography

The Silk Road gives form to a long-held desire to understand the connections of antiquity, how people have migrated over time, and how cultures have interacted over great distances. It is commonly conceived as a historical bridge between the East and the West, one that stretches across from Japan to the Mediterranean. It is portrayed as a network of routes, with its overland corridors reaching as far north as Siberia and southward down through India and Iran. The addition of a maritime route further stretches the story across the Indian Ocean and up through the East and South China Seas. Widespread international recognition of the Silk Road brand means that travel guides and television documentaries frequently introduce Damascus, Palmyra, Xi'an, and Nara as gateways or stopovers in a story of long-distance trade spanning millennia. But for reasons explained in this chapter, this expansive conceptualization of the Silk Road evolved over many decades, and really gained an international following only from the late 1980s onward.[1] To explore this process, the chapter constructs a biography of the Silk Road as a geocultural imaginary of the modern era. There are a number of benefits of this approach. First, it reveals how the concept has traveled and circulated over the decades. Second, it traces how such a narrative of history oriented around long-distance connectivity has evolved and expanded, drawing in ever-more locations and regions, time frames and themes. Third, it enables us to identify the various ideas and values that have come to be associated with the Silk Road over this period. But the analytical threads explored here also show how the concept has become entangled with, and arises from, processes of international diplomacy, geopolitics and imperial ambition, and particular forms of state cultural governance.

To suggest that there is widespread consensus in how the parameters of the Silk Road are defined today would miss the ambiguities, the loosely made assertions, and the very different takes on Silk Road history that remain in play. Since the mid-1990s there has been a proliferation of Silk Road publications, and a survey of this literature reveals contrasting ideas about what the focus should be. For some it remains the diffusion of religions and ideas within Central Asia at particular moments in time. For others the Silk Road is a much more expansive story told through the waxing and waning of overland and maritime trade routes, and the rise and fall of great empires. Elsewhere, the story gravitates around the production and circulation of specific commodities, with silk, porcelain, spices, and carpets among the items commonly used to anchor the narrative. Evidence of silk found in Egypt in the second millennium BCE and the presence of nomadic groups across the Eurasian Steppe give weight to arguments for stretching the timeline much further back than that proposed by Ferdinand von Richthofen in 1877.[2] More recent research on the spread of Islam across the Asian continent also folds in new understandings and ideas about networks of connectivity, taking the Silk Road up to and beyond the arrival of European naval power at the end of the fifteenth century.[3]

It is important to note, then, that the Silk Road remains an unstable concept. In the first instance, instability lies in the identification of its spatial and temporal boundaries. The second, and perhaps more important, instability surrounds the validity of the concept itself. At one end of the table sit those who approach it as an empirically verifiable history of long-distance trade routes and connections, a story of premodern globalization that we are yet to fully comprehend and that requires further investigation.[4] At the other end sit those who regard the Silk Road as a phantasmatic history, wherein complex and disconnected Eurasian pasts are subsumed and pacified in a singular grand narrative. From this latter perspective, the past is romanticized, and the concept of the Silk Road even creates analytical falsehoods. My aim, then, is not to try to resolve questions of validity and the merits of a Silk Road history, or the debate over its scale or scope. Rather, the task here is to reveal some of the ways in which the concept has evolved as it has been used and co-opted across a multitude of contexts. Part of the discussion focuses on the steady accumulation of knowledge over the course of the twentieth century concerning deep histories of transregional connectivity, and I revisit this in later chapters. But the primary intention here is to move beyond the worlds of scholarship and research and examine how the idea has circulated across a number of sectors—tourism, film, literature, international relations, museums, and cultural policy—and across various national and international contexts. In offering a road map for what follows, a number of pertinent themes emerge. First, as a narrative of historical connectivity with origins in the discovery of antiquities in the late nineteenth century, we learn how the Silk Road lies outside the pathways of historiography that dominated the twentieth century. As noted in the previous chapter, fields such as archaeology have long been enmeshed in the politics of nation building and have provided the historical "roots" around which ideas of national identities and cultural nationalisms have been molded and fashioned. In contrast to India, Cambodia, Greece, and elsewhere, the archaeology and philology that gave birth to Silk Road studies were not appropriated and deployed to build an imagined community for the modern era. But as we will see, this "golden age" of Silk Road research in Central Asia did not include sites in Syria, Iran, Iraq, or the Indian subcontinent, many of which are now regarded as Silk Road stopovers. Instead, these are located in regions where the politics of historiography followed the more familiar trajectories of European colonialism and indigenous nationalisms. What becomes evident here, then, is that a narrative of long-distance, cross-cultural connectivity sat on the margins of twentieth-century statecraft practices in Eurasia and their ideologies of historiography and heritage making.

From this emerges a realization that the Silk Road's evolution has been inseparable from the major political events that shaped twentieth-century Eurasia. The concept was born out of geopolitics and imperial ambition, and its contours have been fashioned by the formation of the Soviet Union, two world wars, decolonization, and the Cold War. But where such contexts gave life to finding the routes of shared histories, it required the political and economic reforms of 1980s China, as well as the collapse of the Soviet Union, for the Silk Road to reach the scale and level of international visibility we see today. The identification of national and international politics in the biography of the Silk Road begs the question of how this vision of the past reflects the anxieties, ambitions, and ideals of the present. Accordingly, it will become apparent that the Silk Road has stood as a metaphor for peaceful exchange and coprosperity, adventure, civilizational grandeur, and cultural transmission as it gained a foothold in academia, the international tourism industry, popular culture, international relations, and intergovernmental policy.

The attentive reader will notice a fluctuation between the designation of the Silk Road as both singular and plural. An ambiguity in terminology has been a recurring theme of the past few decades, as geographies have expanded and the idea of a separate Maritime Silk Road gains currency. It is now more common to use the plural, but the singular remains a valid means for signaling the concept itself. The addition of an *s* typically signifies the designation of multiple routes and a recognition that interpretations of the term have continued to evolve. The chapters that follow adopt this distinction.

Constructing the Antiquities of Eurasia in the Politics of Empire

The story of the Silk Road as a geocultural imaginary of the modern era has its origins in the imperial rivalries of late nineteenth-century Central Asia. At the beginning of the century, Russia was on course to become the preeminent land power in Eurasia. Britain controlled Asia's sea lanes and by the end of the 1700s had established the maritime networks required to influence global trade. On land, however, Tsarist Russia continued to expand eastward and, despite being continually challenged by its European counterparts, exerted its influence over vast tracts of land across the continent right through to the early years of the twentieth century. The invasion of Egypt by Napoleon in 1798 stoked fears in London that a Franco-Russian alliance would expand eastward toward Persia and bear down on Britain's prize possession, India. In Asia, Britain's position was undermined by its "geographically and strategically unsatisfactory frontiers," as Ingram puts it, and in the early nineteenth century the threat posed by Russia seemed to only increase as treaties signed in Turkmenchay and Adrianople in 1828 and 1829, respectively, signaled the possibility of Russian protectorates in Persia and Turkey.[5] London and Calcutta were primarily concerned that such a shift in power in western Asia could trigger unrest and rebellion in India, and thus a possible catastrophic bankruptcy of the Government of India. Their response was to build a buffer zone spanning Persia, Turkey, and Afghanistan, in the hope of forestalling Russia's expansion.[6] A British strategy to maintain stable trade and diplomatic relations with the states lying along India's northwestern frontier and an attempt to transform Afghanistan into a client state collapsed dramatically, leading to defeat in the First Anglo-Afghan War of 1839–1842. A region broadly equivalent to modern-day Kazakhstan, the Kazakh Khanate had come under Russian rule, with military personnel and merchant traders attempting to control the conditions of trade across Central Asia. Indeed, despite suffering a defeat in the Crimean War of 1853–1856, Russia was looking to take advantage of a receding Ottoman Empire and maintain access across the all-important Black Sea. By the 1860s a clearer demarcation between British and Russian zones of influence had been established. In the case of Afghanistan, decades of treaties and agreements ensured that the country was continually carved up between the two imperial powers. Across the region, however, resistance from regional powers meant that borders and boundaries remained vague and porous, altering with each new declaration or incursion.

To counter Britain's dominance in sea power, Russia embarked on an ambitious rail-building program from the 1860s onward. Edward Ames documents how key cities were connected with regions undergoing rapid industrialization, with additional lines laid north and south of the Volga and through the Caucasus. With post–Crimean War Russian expansion plans leading to the capture of Tashkent and Samarkand and their surrounding regions, ambitions turned to extending rail lines into Central Asia and beyond. But in this golden era of railway building, Alexander II was far from alone in understanding the strategic value of reaching China by rail. The rapacious demand for coal created by modern warfare and industrialization had created a new breed of explorer with geological expertise. Britain's success in the Opium Wars opened China to foreign exploration for coal and mineral deposits. Tamara Chin argues that Captain T. T. Blaikston was among those who brought a new era of science to the country, creating the prospect of vast riches for those able to capitalize on its abundant resources. With the first transcontinental railway completed in the United States in 1869, American corporations joined their European counterparts in pursuing the new economic possibilities posed by China. Together they sponsored expeditions to survey the geology of China's central and western regions, in the hope of identifying likely mineral reserves and rail routes suitable for transporting the intensely heavy resources over great distances.

The deep suspicion Britain and Russia held for each other throughout this period gave additional impetus to the exploration and surveying of remote lands. Governments on both sides commissioned expeditions to what is today Afghanistan, Inner Mongolia, Tibet, and Xinjiang, areas where few, if any, Western travelers had ever visited. On the Russian side, Nikolai Przhevalsky, a self-taught botanist and zoologist, gained fame through four long and arduous trips as an army officer gathering political, geological, and military intelligence. And it is here that the famed Ferdinand von Richthofen enters the story. Born in 1833, Richthofen developed an interest in geography and geology while on trips to the mountains of southern Europe. After participating in a Prussian expedition to Southeast Asia between 1860 and 1862, he moved to the American West to work as a geologist in the lucrative mining sector for six years. From there, he moved on to East Asia, undertaking a series of surveys over a four-year period. Chin argues that Richthofen took particular inspiration from Blaikston's excursions and British commercial geography techniques. To conduct his surveys Richthofen obtained sponsorship from the Bank of California, which funded his first year and the Anglo-American Shanghai Chamber of Commerce, which supported three further years.[7] On his return to Germany, he took up an academic position as a geologist and completed the first part of *China*, a publication that would appear over five volumes between 1877 and 1912. As Waugh notes, in the first volume Richthofen lays out his commitment to understanding the region as a geographer, stating that an account of the physical landscape must be accompanied by an explication of the dynamics of human settlement. Focusing primarily on the theme of trade networks established during the Han dynasty (206 BCE–220 CE) and the shifting significance of silk as a luxury and commodity, Richthofen combined his fieldwork with the work of others to demonstrate, as Waugh puts it, "the history of geographic knowledge in the West with regard to China and conversely, in China with regard to the West."[8] It is here that he introduces the term *Seidenstrasse* (Silk Road).

Tamara Chin offers a wonderfully detailed account of the factors that determined Richthofen's cartographic choices, noting that the significance of his approach stemmed, in part, from the integration of Chinese sources with maps and writings produced by Marinus of Tyre and Claudius Ptolemy from the first and second century, respectively.[9] Ptolemy's *Geography* documented the accounts of long-distance travelers in order to create the most comprehensive maps of the Greco-Roman world, and upon its rediscovery in the fifteenth century, the text continued to influence exploration and cartography for centuries. But Richthofen's cartography also incorporated the account of Marco Polo and, most innovatively, Chinese sources depicting histories of silk exports. To distinguish the different routes taken around the Taklamakan Desert, he used red ink to plot European sources and drew a blue line to depict a southern route used by Chinese traders. To extend his Silk Route west of Balkh, in present-day Afghanistan, and onward to Europe, Richthofen reproduced the route plotted by his German rival, Heinrich Kiepert. Chin reminds us that these cartographic exercises were, in part, oriented around the calculation of distance. The late nineteenth-century versions of Kiepert and Richthofen thus maintained latitudinal consistency with their predecessors, as they connected Europe to Asia in the shortest possible manner. This meant that they reaffirmed the routes of ancient trade passing through the north of Iran, the Caucasus, and on to southern Europe. In the 1880s, the identification of accurate distance and direct routes carried the impetus of a transcontinental rail line. Richthofen's efforts to produce a map for prospective rail construction thus involved assessing distances and terrain. Routes plotted for the stretch between Balkh and Xi'an would also provide a template for the historical Silk Road (fig. 2.1). Interestingly, however, his 1877 article and book featured both the singular *Seidenstrasse* and its plural *Seidenstrassen*. But as Chin points out, English versions of the article in the *Geographical Magazine* and *Popular Science Monthly*, published in Britain and the United States, respectively, a year later, referred to only a single Silk Road.[10] The distillation of complexity and ambiguity through translation and transposition had begun.

With Russian, British, and Qing officials all looking to expand the lands they had already wrested control over, Chin suggests that Richthofen's route also reflected the complex territorial politics of the period. All parties understood the strategic importance of rail: long-distance lines provided revenue streams through construction and concessions, gave access to coal deposits, and helped secure trade and strategic interests. As we will see shortly, Richthofen's student Sven Hedin helped popularize the Silk Road concept in the 1930s. But in the reports of their respective trips undertaken during the late nineteenth century, it is evident that both men tailored their geological and geohistorical knowledge to the possibilities of large-scale infrastructure investments and the political and commercial rewards they could deliver.[11] By 1888 Russia had completed the first phase of the Trans-Caspian Railway, connecting Turkmenbashi (then Krasnovodsk) on the Caspian Sea with Samarkand. A decade later it reached Tashkent and Andijan in present-day Uzbekistan.[12] In 1891 work began on the eastern end of a line that eventually became the Trans-Siberian. An extraordinarily ambitious project designed to connect Moscow with the strategic port city of

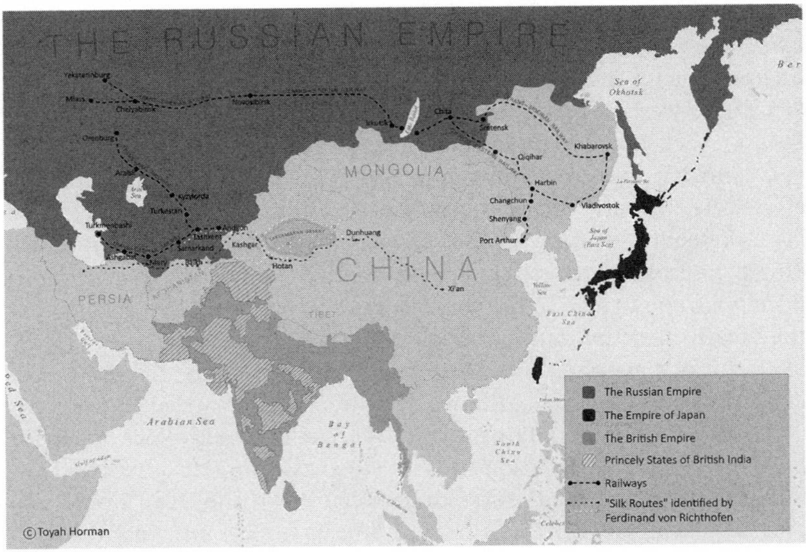

2.1 The Great Game (c. 1900) and Silk Roads of Ferdinand von Richthofen. Courtesy of Toyah Horman.

Vladivostok in the east, the Trans-Siberian Railway was constructed in sections at great financial and human cost over a span of nearly twenty years (see fig. 2.1).[13] Writing in the early 1930s, W. E. Wheeler outlined the importance of the two lines, together with the Turkestan-Siberia and Chinese Eastern lines, for securing economic control and the movement of large quantities of troops and military hardware into vulnerable areas. Emphasizing their broader strategic significance, he stated that "the importance of the Turk-Sib is implied in the industrialization of the Soviet state and the military defense of that industrialisation."[14]

Metropolitan centers in Europe, India, Japan, and the United States were increasingly aware that Central Asia was becoming "the strategic cockpit of the continent."[15] In Britain, George Curzon, later Lord Curzon, published his highly influential *Russia in Central Asia* in 1889, a manuscript based on extensive travel in the region. Inspired by Curzon's analysis, Halford J. Mackinder presented the article "The Geographical Pivot of History" to the Royal Geographical Society in 1904.[16] Mackinder argued that Russia's infrastructure of inland railways marked a shift in the balance of power over Britain's naval resources.[17] He thus concluded that the politics of the Eurasian heartlands would shape the future of world affairs. This idea of a strategically critical "pivot area" influenced Western military thinking and geopolitical theory for decades to come. But one concept, more than any other, captured the imagination of this period. In a chapter subtitle in his 1851 *History of the War in Afghanistan*, the former army officer John William Kaye made reference to "the great game."[18] Seymour Becker has traced the usage of the term through the nineteenth century, suggesting that it came to be more widely associated with the novel by Rudyard Kipling, *Kim*, not published until 1901. As we will see later, the expression gained its fullest notoriety during the Soviet-Afghan War of the 1970s.

In the West the historicization of the Great Game has largely focused on the rivalry between Russia and Britain, with the term more commonly associated with the latter.[19] Less attention has been given to the ambitions of Qing-dynasty China or Japan (fig. 2.2). The construction of rail lines across East Asia brought Russia into an uneasy proximity with Japan. At a time when the Japanese were developing their own expansionist plans, the reaches of European empire building were viewed with considerable consternation. With leaders of the Meiji period wanting to avoid the humiliation of their regional neighbors, military ambitions turned to the Korean Peninsula and Manchuria. But as Esenbel points out, the internationalization of Japanese modernity at this time also meant that missions were dispatched farther afield for trade and cultural purposes. Official trips to the Qajar monarchy in Iran and the Ottoman capital of Constantinople, for example, helped establish commercial and diplomatic ties with the major Muslim polities of West Asia.[20] With such trips, Japan, like the British, sought to build a buffer zone against imperial ambition. Once again, investments were also made in the surveying of great areas of Central Asia, China, and regions to the north in anticipation of military ventures.[21] Competition between Japan and Russia for control over Manchuria led to the conflict of 1904–1905. Japan had expanded its army presence in the region in the years following the First Sino-Japanese War of 1894–1895, and the defeat of a European power a decade later marked a turning point in Asia's political affairs and created a new platform for regional alliances based on culture and history, a theme I return to shortly.

Nile Green has considered such events through the various forms of written text produced during this period. Accordingly, he suggests, those who traveled across the region "both participated in and documented the larger transformations" of the period.[22] Over the course of the nineteenth century, explorers, spies, and army personnel returned home with stories of the ancient ruins, monasteries, and lost cities of Central Asia. Such details, however, were often presented as anecdotal

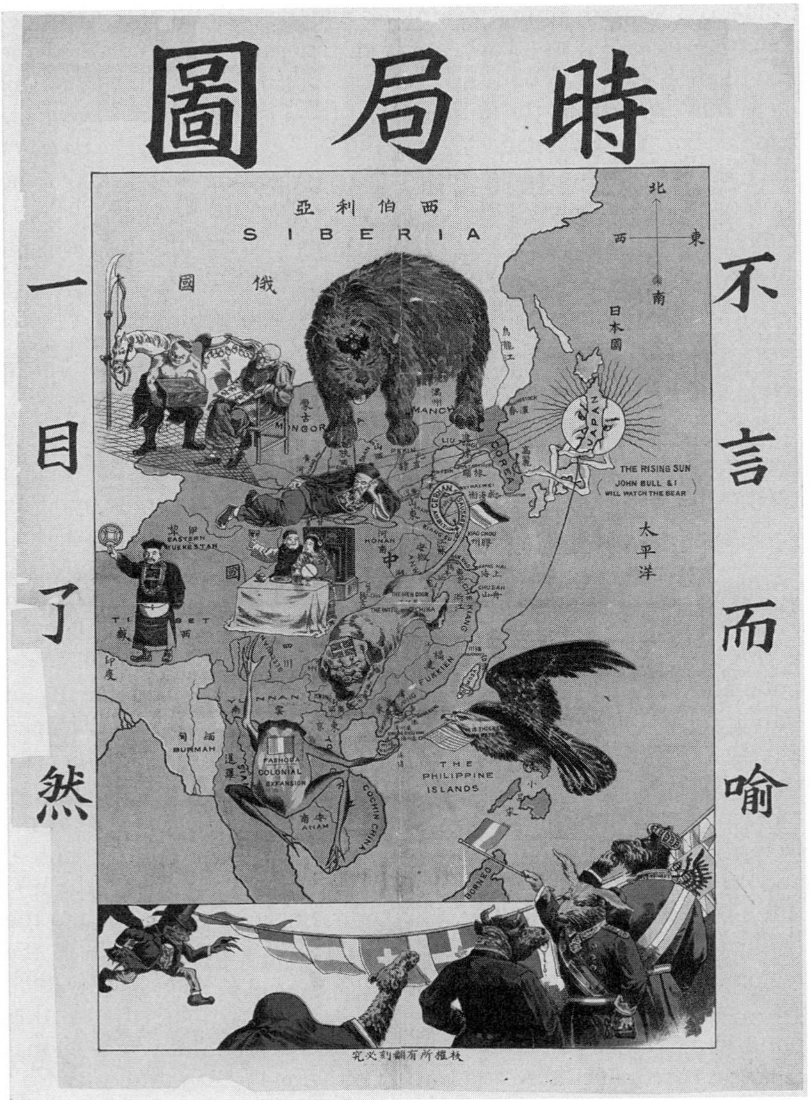

2.2 "The Situation in the Far East," by Tse Tsan-tai (1872–1939).

comments or the stuff of legends in personal diaries and military intelligence reports. During his trips to Tibet in 1871 and 1876, Nikolai Przhevalsky, for example, noted staying in monasteries and stumbling across abandoned structures and settlements. Likewise, the Russian botanist Albert Regel recounted discovering the extensive ruins of the ancient Uyghur capital Karakhoja (Gaochang), and in that same year, 1879, a Hungarian surveying expedition arrived at the caves of Dunhuang on the southern edge of the Taklamakan Desert.[23] For the British, intelligence gathering involved risky and hazardous excursions from India. As Peter Hopkirk notes in his acclaimed volume *Foreign Devils on the Silk Road*, on the international race for archaeological remains in Central Asia, in the 1870s and 1880s the British often chose to send local Indian clerks on reconnaissance trips. Equipped with instructions and the basic materials for conducting surveys of the topography and locations of Russian military camps, they would return with stories of lost cities and priceless treasures.[24] Such accounts filtered back to the metropolitan centers of Europe but invariably garnered little excitement among an archaeological profession preoccupied with unearthing the secrets of the great civilizations of the Mediterranean, Egypt, and lands associated with the Bible.

The discovery of a manuscript in 1890 in Kucha, in present-day Xinjiang, altered this situation, opening the window to previously unimagined histories of civilizational contact and premodern mobility. The Bower Manuscript, named after the British officer Hamilton Bower, who purchased it from a team of local treasure hunters, comprised seven parts. Considered to date back to the fourth or fifth century, the manuscript's Sanskrit-language Gupta script provided the first evidence of contact between India, China, and Central Asia, and also revealed the depth of the region's literary traditions and ideas about medicine. Published translations presented scholars and archaeologists across Europe with a raft of new questions concerning the diffusion of Buddhism and languages and the historical scope of cross-cultural interaction more broadly. A rush to discover additional manuscripts soon followed. Throughout the 1890s teams from Britain, France, and Russia all set off in search of ancient texts, heralding a new age of collecting and international archaeology. The study of manuscripts and mural fragments by scholars and museums in St. Petersburg led to the Russian Academy of Sciences' organization of an archaeological expedition to Eastern Turkestan in 1898. Led by Dimitri Klementz, the trip returned with reports and photographs of ruins located on the edge of the Gobi Desert, near Turfan.[25] The prospect of discovering lost cities also inspired Sven

Hedin to venture deep into the inhospitable terrain of the Taklamakan Desert. Hopkirk complements Hedin's untrained eye in documenting the Indian, Greek, Persian, and Gandharan influences in the iconography he found.[26] During his second expedition, which ran from 1899 to 1902, Hedin also claimed to be the first European to visit the Buddhist stupas, statues, and carved reliefs of Loulan, a garrison town that had flourished until the fourth century CE on the eastern side of the Taklamakan Desert.

The British had also accumulated a sizable collection of Central Asian artifacts in Calcutta. But it would take the appointment of Lord Curzon as viceroy in 1898 to ensure a long-term commitment to archaeology beyond India's frontiers. Curzon publicly lamented the Indian government's lack of care and investment in the country's antiquities and religious monuments. To address the issue, he appointed himself director of antiquities and established the Archaeological Survey of India. In April 1899, Curzon met Aurel Stein, principal of the Oriental College, Lahore, and a keen historian of the region. Stein was in search of expedition funding, and Curzon's promises of support yielded a grant some months later. Stein promptly set off on a trip to Chinese Turkestan, an area that by this time had become politically sensitive for the British to visit. Born and initially educated in Budapest, Stein studied Persian and Sanskrit at university. Postdoctoral appointments took him to England, and after a period of military service back in Hungary, he set sail for India at the age of twenty-five. There he put to good use the surveying and mapmaking skills he acquired in the military. But it was the years spent studying coins, artifacts, and texts in the museums and university libraries of England and Central Europe that cemented his lifelong passion for Asia's history and cultures of antiquity. Before his death in Kabul in 1943, Stein undertook four lengthy expeditions across Central Asia as well as a series of trips through Iran and other parts of the Middle East. His maiden trip of 1900 began in Srinagar, and he reached Kashgar some months later. There he entered a world afflicted by decades of political upheaval. On the back of defeating the Tajik ruler Yaqub Beg in the 1870s, since 1884 Peking had declared Xinjiang a province of the Qing dynasty. By the time Stein arrived in 1900, however, the level of Russian influence had grown considerably. The Boxer Rebellion had broken out, and the Russian army was in the process of invading Manchuria. Nicholas Petrovsky, stationed in Kashgar since 1882 and acting in the capacity of consular to Peking, exerted his influence over the affairs of the region. Like Curzon, Petrovsky was keenly interested in history and languages, lending his support to Russia's growing archaeological programs. Vsevolod Ivanovich Roborovskiy had conducted pioneering archaeological work in the Turfan region between 1893 and 1895.[27] Dimitri Klementz's 1898 excursion also shed light on the contents of 130 Buddhist cave temples some distance north of Kashgar in Turfan. As Stein ventured out from Kashgar, he thus avoided areas north of the Taklamakan under Russian control and instead headed south toward Khotan.[28]

In such examples we begin to see how the archaeological excursions of this period reflected the political frontiers and lines of control being drawn up across the region. As figure 2.3 illustrates, Russian archaeological expeditions were predominantly undertaken in the area north of the Taklamakan Desert, whereas the key archaeological sites for the British reflected their route of access from India in the south and their safe passage along the southern rim of the desert. Not surprisingly, archaeology was fast becoming an arena of competition and rivalry. But this meant that it also furnished opportunities for international cooperation and collaboration, even across the fault lines of imperial rivalry. Petrovsky granted permission for Stein to return home to Europe with his cargo of twelve boxes of artifacts and manuscripts via the Trans-Caspian Railway. But as Stein explored alone, teams from Russia and Germany set about planning joint archaeological expeditions.[29]

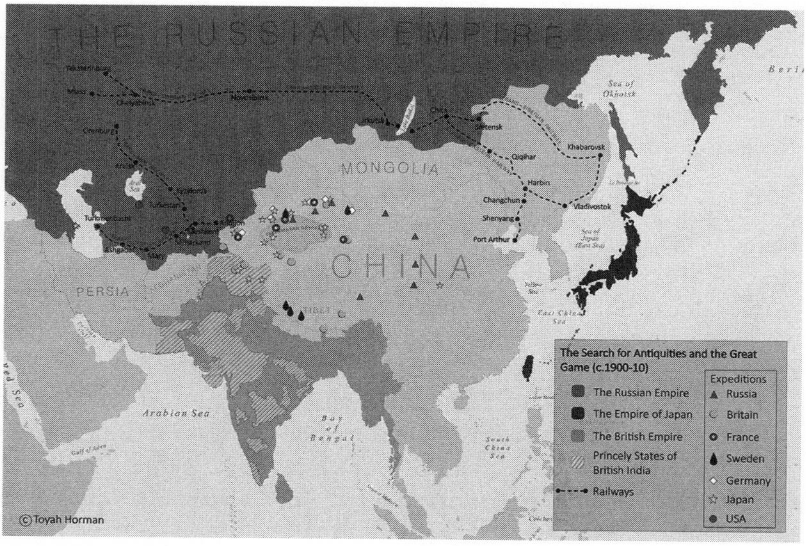

2.3 The search for antiquities and the Great Game (c. 1900–1910). Courtesy of Toyah Horman.

In 1899, Dimitri Klementz visited the Berlin Ethnological Museum en route to Rome to attend the International Congress of Orientalists. The museum's director Albert Grünwedel excitedly confirmed Indian and Iranian influences in the Buddhist frescoes. A Russo-German collaboration was proposed, and the first resultant expedition returned from Turfan with nearly fifty chests of artifacts in 1903. A year later, Albert von Le Coq led a second trip, one that involved the shipment of a third of a ton of items, including frescoes and wall carvings extracted by saws. At that point the Trans-Caspian Railway had reached as far as Tashkent, greatly easing passage into Central Asia. This new iron Silk Road increased both speed of travel and the weight of cargo able to be carried back and forth. Archaeological trips undertaken in the years before World War I helped Germany maintain its reputation as a leading force in Oriental studies. By 1912, the Berlin Turfan collection housed more than thirty thousand manuscript fragments in readiness for research and public display. But as Strauch points out, these collaborations also helped the Russians catch up with Britain in the prestige stakes and obtain expedition funding when their own government was reluctant to commit.[30]

The International Congresses of Rome, Paris (1897), and Hamburg (1904) enabled scholars from across Europe and the United States to share ideas and findings. Accounts of Silk Road histories by Hansen, Wood, and others demonstrate how the research conducted in this period shed new light on the extent of connections between Europe and Asia stretching back millennia.[31] As Esenbel notes, the insights gleaned from Central Asia helped forge new understandings of world history: "In European scholarship, the Silk Road signified the exciting discovery of cosmopolitan encounters and plurality of national identities of an indigenous demography that incorporated the historical interconnections between nomads, traders, and settled peoples, independent of the neighboring Russian and Chinese empires. . . . [It] represented an extraordinary 'international' history of the descendants of the Greeks, Nestorian Christians, Sogdian merchants, Turkic and Mongol nomads, Uyghur kingdoms, Chinese travelers, and Indian pilgrims."[32]

Today, the visit by Aurel Stein to a network of grottoes a few miles outside Dunhuang in 1907 is widely regarded as the pinnacle of this brief but golden era of archaeology and philology. At the beginning of the twentieth century, Dunhuang remained a remote location, but at its height it acted as a crossroads on the two principal trade routes bordering the Taklamakan, and was an important stop on the north-south road connecting Mongolia with Tibet. A witness to Mongol invasions and Muslim conquests, the site was also a refuge for Buddhist monks for more than a thousand years between the fourth and fourteenth centuries. During this period an estimated forty-five thousand murals were painted across a landscape comprising around five hundred grottoes (fig. 2.4). In the now-infamous story of Stein's discovery, in excess of fifty thousand manuscripts were found hidden behind a bricked-up cave entrance. The grottoes were under the careful watch of a local Taoist monk, Wang Yuanlu, and after some careful negotiation, Stein persuaded Wang to allow a small sample of items to be carted off to England. Meyer and Brysac describe Stein's haul of twelve cases in the following terms: "Paintings on silk, embroideries, sculptures, and, most importantly, more than a thousand ancient manuscripts written not only in Chinese but also in Tibetan, Tangut, Sanskrit, Turkish and a scattering of other, obscurer languages. The crown jewel was the world's oldest known printed book, dated May 11, 868, the Diamond Sutra, a popular spiritual text. All this, Stein proudly wrote to his friend P.S. Allen, cost the British taxpayer a mere 130GBP, the sum he donated to Wang for the upkeep of the site."[33]

In the years that followed, Dunhuang was visited by teams from France, Japan, and Russia. Among the artifacts shipped to Paris by Paul

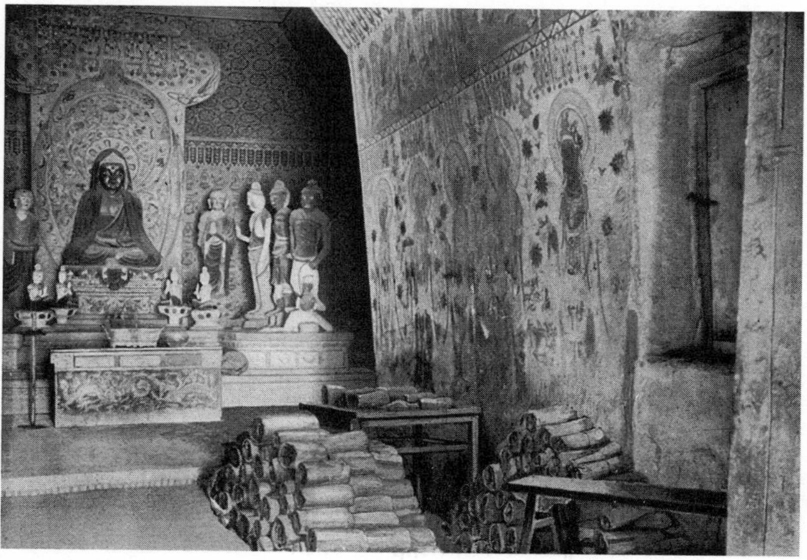

2.4 Dunhuang Library Cave. Photography by Aurel Stein, c. 1907. Source: The British Library Board.

Pelliot—the highly talented Sinologist employed by the École française d'Extrême-Orient—were fifty pieces of sculpture. Hundreds of manuscripts were acquired for museums in St. Petersburg by the acclaimed Russian Orientalist Sergei Oldenburg. Stein returned to the region as part of a three-year expedition that began in 1913 and included an exploration of the Mongol Steppe, the Pamir Mountains, and Iran. The discoveries made at Dunhuang cemented the idea of Central Asia as an important center of civilizational exchange and cultural production. Stein was fascinated by the connections to India, and further interpretations of his finds using the writings of Marco Polo and Ptolemy represented the first forays into a story of long-distance trade stretching from the Mediterranean to East Asia, one that would eventually underpin the concept of the Silk Road. But while European researchers were engaged in building this narrative of east-west trade and cultural transmission, Central Asian antiquity held quite different meanings for Japanese intellectuals.

On a trip to London in 1901, the Japanese monk and scholar Ōtani Kozui learned of the expeditions conducted by Hedin and Stein. Ōtani would later become the chief abbot of one of Kyoto's most important temples, and as a scholar of Buddhism he wanted to learn about the great finds made by European scholars in Central Asia.[34] In the aftermath of the Meiji Restoration of 1868, Buddhism was in retreat, with Shinto installed as the state religion. Ōtani's travels to Europe as well as other parts of Asia were thus driven by a desire to comprehend the compatibility of religion and modernity on the one hand and the possibilities of using Buddhism as a platform for a pan-Asian revival on the other. As Galambos has demonstrated across a series of fascinating publications, Ōtani traced this transnational, shared heritage by conducting multiple visits to India, China, and other parts of Asia.[35] Of significance here are his three trips to Central Asia undertaken between 1902 and 1914.

Ōtani's expeditions were designed to identify the "Buddhist culture and its cultural and artistic artifacts," which had been forgotten or buried in regions where Islam had become prevalent.[36] After an initial trip spanning 1902–1904 to the ruins and caves of Kucha, Ōtani began work on a villa in Kobe designed to showcase and study the sacred texts and artifacts retrieved by the team. A close relationship with the well-traveled architect Ito Chuto led to a building that drew on Mughal and Persian architecture, as well as Japanese elements, to give a sense of a shared Asian tradition.[37] During a follow-up expedition to Xinjiang and Mongolia in 1908, the team split into two groups in search of historically significant sites. It was here that Ōtani's expedition entered the official British records and thus the story of Hopkirk's "foreign devils."[38] The British consul in Kashgar suspected two members of spying for the Japanese government. On all sides of the Great Game, men gathering military intelligence had been masquerading as archaeologists. Having left their collected items at the British consulate to head south to India, one of the team members was refused entry back into China. Galambos has compared archive sources in Japan and the United Kingdom to argue that there is little evidence to support the suspicions of the British.[39] On a third trip, spanning 1910–1914, team members gathered a wealth of items—including 250 manuscript scrolls from Dunhuang—transporting them back to Japan via Siberia on the Trans-Siberian Railway. Ōtani extensively publicized the results of his three expeditions, communicating with scholars and associations across Europe. But he also endeavored to include Japanese Buddhist students and researchers in an effort to enhance their international reputations.[40] For Galambos, then, the importance of the team's search for ancient Buddhist relics, along with Ōtani's larger vision of a pan-Asian cultural past, was twofold: "On the one hand, they tried to position themselves as part of the tradition of European exploration of Central Asia, which to some extent reflects Japan's contemporary aspirations to align itself with leading colonial powers. This European colonial element shines through in all narratives associated with the expeditions. On the other hand, they also claimed a connection with the medieval Buddhist pilgrimages through the Western Regions."[41]

Japan's victory over Russia represented a landmark moment for those invested in a pan-Asian ideology. The movement had its roots in the humiliation of China in the Opium Wars of the mid-nineteenth century, and the defeat of a major European power gave significant impetus to sentiments of nationalism across Asia. The confidence and resilience accumulated during the Meiji period created a strong platform for the emergence of pan-Asianism in Japan. Ōtani's contributions were both significant and timely given that, as Saaler and Szpilman note, the concept was only beginning to appear in intellectual discourse in the early 1910s. Pan-Asianist writers and other critics of European colonialism stressed regional solidarity and by the 1920s such groups were holding conferences in China, India, and Japan. A visit by the celebrated poet Rabindranath Tagore to Japan in 1916 helped cement ties with an Indian intellectual elite increasingly disillusioned with colonial rule.

Nationalist and regionalist identities needed their historical roots, and in India decades of archaeology, and particularly the work of Alexander Cunningham on the seventh-century Chinese traveler Xuanzang, demonstrated the long-standing ties between India and China. The spread of Buddhism also gave weight to an intellectual position that foregrounded harmonious, precolonial intra-Asian interactions. Tansen Sen thus concludes that ideas about contact zones between "Asian civilizations" served as the foundations for building ideas of cultural and political regionalisms.[42] Of course, all this would be facilitated by the new infrastructures of modernity. Rail and steamships carried Japanese visitors to the most sacred sites of Buddhism in India, most notably Bodh Gaya. Mark Ravinder Frost also points to the role played by the region's network of museums, which displayed the items collected by travelers and scholars. In Calcutta, the India Museum was home to an extensive collection of artifacts from across the Central Asia region.[43] Crucially, then, the pan-Asianism that emerged in the early decades of the twentieth century constituted a geocultural imagination incorporating the regions, histories, cultural interactions, and religions that today come under the umbrella of the Silk Road. It advanced new narratives about civilization in Asia that were entirely disconnected from the canonical religions or major events of European history.[44] At a political level, such a reading of the past refuted Eurocentric visions of the world order, which placed Western civilization at the center of universalist discourses of progress and enlightenment.[45]

It is important to remember that this narrative of history emerged in dialogue with the European archaeology, philology, and manuscript studies outlined earlier. Ōtani and others embraced scholarship that foregrounded the Europe-Asia nexus. Interestingly, however, the combination of these geocultural imaginations—pan-Asianism and east-west exchange—also brought Japan into dialogue with the Ottoman Empire. For pan-Islamic intellectuals in search of a resolution to an expanding Tsarist Russia, Japan served as an inspirational model for Islamic modernity.[46] Dündar demonstrates how this came to be expressed across the Turkic world via poetry and literature.[47] But with pan-Asianism underpinning the imperial ambitions of the Japanese ultranationalist right by the 1920s, anti-Russian sentiment in both Turkey and Japan fueled ambitions for a more united Asia. Esenbel argues that this situation encouraged a number of Turkish intellectuals to turn to geocultural histories as the connective tissue. Interestingly, ties between the two countries continue today via a number of Japanese-Turkish scholarly collaborations in Silk Road studies.[48]

The key assertion here, then, is that the geopolitics of nineteenth-century Central Asia, an arena of competing interests and imperial rivalries, constituted a milieu in which particular ideas about culture and history were fashioned. In its emergent infrastructures and fluid, ambiguous notions of space and territory, the Great Game fomented a historicization of the region oriented around long-distance connectivity rather than the cultural-historical roots of a particular people or civilization. Unlike the fields of Egyptology or Southeast Asian archaeology, there were no great temples or mausoleums for nineteenth-century scholars to build narratives of kingdoms or premodern empires. By implication, the discourses of archaeology or philology in Central Asia did not revolve around attempts to designate the territorial reach of these kingdoms or how they might correspond with the domains of their nineteenth-century descendants. Instead, the interpretative frame was long-distance connectivity, cultural transmission, and a dialogue between religions. Equally significant, and again, unlike elsewhere in Asia or the Middle East, this historical research was not caught up in a colonial apparatus seeking to build the imagined communities and geographical boundaries of modern nation-states.[49] The caves of Dunhuang and the lost cities of Turfan were remote outposts to European colonialism and thus were not subject to the types of political appropriation and heritage and memory industries constructed around Ayodhya, Borobudur, Persepolis, or Angkor. The story of Ōtani's expeditions also opens up the important theme of the growing interest in intraregional geocultural relations, which largely emerged on the back of changing domestic political landscapes. As Judith Snodgrass has shown, Buddhism was vital for those Japanese wanting to build international connections—whether they be configured personally or via institutions—and build a political base at home.[50] For the pan-Asianists, Central Asia was a key contact zone connecting countries in East and South Asia. In the decades following World War II, Japanese institutions continued to commemorate such intraregional histories of Buddhist antiquity and art, tying them to a Silk Road concept that was steadily becoming better known around the world.

Interestingly, during the opening decades of the twentieth century, seeds were also being sown for an intraregional maritime historiography. Kwa Chong-Guan makes the point of connecting Paul Pelliot's interests in Central Asia with his commitment to comprehending the region's maritime past. Pelliot was among a number of Sinologists who searched historical records for evidence of how knowledge about the southern seas accumulated in India and China. Such French and Dutch scholarship, he suggests, inspired a generation of Indian historians, mostly Bengalis, to form the Greater India Society in 1926.[51] Members of the society were intrigued to better understand India's maritime ties across the Bay of Bengal and down through Southeast Asia. The diffusion of language and Buddhism and Hinduism was testimony to an Indian civilization carried by merchants and other seafarers over a number of centuries. Again, we will see such ideas continue to develop after World War II, and indeed, as Kwa suggests, such developments, including the Greater India discourse, were important antecedents to today's idea of the Maritime Silk Road.[52]

Inscribing National Pasts

World War I transformed the political landscape of Europe, and, together with the fall of the Ottoman Empire and revolutions of 1917 in Russia, remapped the center-periphery relations of Eurasia. The formation of the Soviet Union in 1922 set in motion changes that would disconnect Central Asia from the West for much of the twentieth century. With key aspects of the Great Game coming to an end, the door began to close for those invested in understanding the historical connections between East and West. Equally significant, for a number of countries and regions the dynamics of historiography were also about to take a different turn. In the academic institutions of Tsarist Russia, archaeology had built a solid intellectual base. Its position relative to the imperial state, the monarchy, and public was secured by regular congresses and the famed museums of St. Petersburg and Moscow.[53] The rise of political Marxism, however, brought wholesale changes to the humanities and social sciences, transforming both what was studied and how it was interpreted. With a new generation of scholars entering the fields of archaeology and anthropology, by the 1920s previous typologies of culture and history were abandoned and discredited in favor of a Marxist conception of historical "stages" and their socioeconomic formations.[54] To accommodate these intellectual shifts archaeology was renamed "the history of material culture."[55] The historical centers of scholarship remained in Moscow and St. Petersburg (then Leningrad), and as the Soviet Union expanded over the course of the 1930s, they trained scholars for positions in the newly established National Academies of Sciences of the republics in the Caucasus. As Dolukhanov explains, Russian archaeologists undertook the bulk of the archaeology conducted in Central

Asia in the years leading up to World War II.[56] But the nature of the research they conducted altered once again in response to the turmoil of the period. Bulkin, Klejn, and Lebedev describes "an especially rapid increase in knowledge of the ancient past" across Central Asia, the Caucasus, and Siberia as part of an effort to promote the cultural enhancement and national self-consciousness of groups living in these regions.[57]

Soviet ethnography followed a similar path. On the back of the formal constitution of the Soviet Union in 1922, the Bolsheviks increased efforts to implement a nationalities policy.[58] *Korenizatsiia,* or "putting down roots," as it is commonly translated, was conceived of to fix the wrongs of the Russian Empire by creating a federal union defined by diversity and unity, an ethos encased in a Marxist ideology of modernization and developmental progress.[59] New policies on language, education, and the political representation of non-Russians also extended to prescribing categories of ethnicity and their associated territories. In her fascinating book *Empire of Nations,* Francine Hirsch focuses in on three technologies of governance—map, census, and museum—all of which formed part of a strategy of double assimilation, whereby diverse groups were categorized both into national populations and into wider Soviet society.[60] As Hirsch illustrates, ethnographers played a key role in guiding the implementation of such policies. Oksana Sarkisova also traces the processes by which this ideology was transposed into a series of state-funded documentaries. Initially given the title of *Kulturfilm,* a term borrowed from Germany, these films were designed to provide audiences—primarily metropolitan Russian—with a visual vocabulary for understanding the Soviet Union as a vast land of immense linguistic and cultural diversity, yet one that spans the Eurasian continent with a single ideology. Filmmakers set off on expeditions throughout the 1920s, frequently collaborating with ethnographers and cartographers. As Sarkisova explains, "*Kulturfilms* gave a tangible form and shape to imaginative concepts such as civilization and backwardness, and highlighted the entanglement of colonizing and modernizing attitudes in the Soviet context."[61] For Central Asia, this involved producing films on both the traditional practices of different cultural groups and the implementation of infrastructure projects, most notably the Turksib Railway (see fig. 2.1). *Krysha Mira,* or *The Roof of the World,* made in 1927, involved a trip to the Pamir Mountains for, what Sarkisova describes as, a "taxidermic" representation of ethnic prototypes based on nomadic or sedentary lifestyles.[62] By contrast, Viktor Turin's much celebrated 1929 film *Turksib* portrayed the transformational power of

Soviet modernization projects, in this case a railway, and their ability to help the remote regions of Central Asia "leap straight into a brave, new, industrial world."[63]

Across the border in Xinjiang, ideas about the region's history were also being reworked on the back of major political change. The above discussion of European archaeology in Chinese Central Asia reached as far as the twilight years of the Qing dynasty, which collapsed in 1912. Throughout the period, foreign travelers and scholars enjoyed the privileges accrued from China's vastly weakened position relative to foreign powers. For those local bureaucrats who came into contact with foreign expeditions, their primary concern was the avoidance of diplomatic incidents. As Justin Jacobs points out, this trumped any concern for questioning, let alone policing, the removal of artifacts.[64] By the late 1920s this situation had changed significantly. Both Stein and Hedin made return visits to Xinjiang shortly after Chiang Kai-shek declared Nanjing the capital of the Republic of China in 1927. In a fascinating account of what he refers to as the "domestic geopolitics" of their two trips, Jacobs describes how they came to be manipulated both by the Nanjing government and by the provincial governor of Xinjiang, Jin Shiren, under the guise of cultural sovereignty. In the face of heavy Soviet interference and unrest from local warlords, Nanjing attempted to retain control over its frontier province. In 1929 Sven Hedin traveled from Europe via the Trans-Siberian Railway, leading a team of five Swedes and two Chinese. Suspicious of their motivations, Jin declined them entry but relented under pressure from Nanjing, which in turn took instruction from powerful academics in Beijing.[65] Nanjing's support for the trip stemmed from a signed agreement that any artifacts found by the team would be deposited in a museum in Nanjing and that the expedition would retain two Chinese members. The trip spanned several years despite the suspicions of Jin. Supporting this Sino-Swedish collaboration served the national unity agenda of the Nationalist Party elite, but as Jacobs reveals, so did shaming Stein's expedition, which set off a year later. Supported by the Fogg Museum of Harvard University, Stein made the strategic mistake of traveling to the region without Chinese partners.[66] The Chinese press thus cast him as a potential looter and threat to cultural sovereignty. His presence in the region also made the government in Nanjing suspect he was undertaking military surveys for the British, in support of a possible move by Jin for greater autonomy. Given the air of mistrust over Stein's motivations, both Jin and Nanjing sought to retain their credibility in the media by publicly declaring their commitment to national unity. Stein left the following year with a cloud over his name, never to return to China.

Evidently, this episode signaled a major transition in the domestic politics of antiquity and how the cultural past of China's frontiers had become part of the discourse of nationalism and aspirations for revival.[67] Elsewhere, Jacobs has traced this process through to the 1940s, with respect to the establishment of the wartime capital Chongqing.[68] He argues that the move gave new symbolic significance to Dunhuang as the political and intellectual elite searched for a new spiritual heartland after an enforced retreat from the Japanese army. As Chinese scholars visited the caves for the first time, they were struck by their grandeur and began absorbing their artworks into ideas of a Han Chinese ancestry. According to Jacobs, Dunhuang's frescoes were reproduced as wartime museum displays, and, by citing the renowned painter Zhang Daqain, he illustrates how deep histories of contact with Europe were framed for a public embattled by conflict with an imperial Japan:

> By closely scrutinizing the features of the people depicted in the Dunhuang frescoes and other Tang paintings, Zhang was able to conclude triumphantly that "the moustache and hair resemble those of western Europeans." The implication was that Tang Chinese "clothing and cultural trappings . . . had once spread all the way to western Europe," such was the strength of ancient China back in the day. Because the Dunhuang murals preserved intact the spirit of those vigorous and admirable Chinese who flourished during the Tang, the fine art connoisseurs who crowded Lanzhou and Chengdu's exhibition halls during 1943–1944 could gaze upon Zhang's lifelike copies and silently intuit the long-lost cure—the cosmopolitan Tang imperial spirit—for China's current malaise.[69]

In 1936 Sven Hedin published an account of his travels in China. An evocative, fluid writer, Hedin's publications reached much wider audiences than those of his counterparts, including Stein. His 1936 work was simply called *The Silk Road*. Published initially in Swedish and translated into German and English, the book was among the texts that helped popularize the term.[70] It had previously featured in book titles, most notably in a volume by Albert Herrmann back in 1910, but in the early 1930s the Silk Road remained a neologism.[71] Herrmann retained Richthofen's narrow definition of the term, tying it primarily to Han dynasty China, and worked toward a more accurate mapping of the trade routes and possible East-West connections of the period.

As I explain in greater detail elsewhere, by the early 1930s adventurers, documentary makers, and writers in publications such as the *National Geographic Magazine* were beginning to use the term, often tying it back to Marco Polo.[72] As Waugh notes, by this time the Silk Road had become associated with the "romantic aura" of great adventure.[73]

To pull a number of threads together, then, Hedin helped popularize a term that had emerged from an episode of exploration and research that took shape within the competition for control over Central Asia. A "golden age" of scholarly exploration in the region in the late nineteenth century had sown the seeds for a grand story of historical connectivity and long-distance cultural transmission. As we have seen, though, the arrival of international conflicts and major political transitions meant that the structures required for funding and enabling such scholarly undertakings either disappeared or shifted in ways that foreclosed the ongoing development of this line of inquiry. In the brief examination of early Soviet archaeology and ethnography, it is evident that both became important resources in the designation of national and territorial categories of culture and formed part of a Soviet governmentality oriented toward modernization and development. In Central Asia this trend continued after World War II via policies for heritage conservation, tourism, and museums. Clearly, Soviet era strategies for producing borders through the alignment of territories with language and carefully prescribed cultural identities represents a situation entirely at odds with the celebration of a Silk Road heritage of flows and cross-cultural transmission we see today. But what is apparent in 1930s Central Asia is the continuation of themes and intersections first established in the late nineteenth century, which would remain present as the Silk Road concept gained international visibility toward the end of the Cold War.

At this point the story moves westward, to a region where researchers and explorers were enmeshed in quite different geopolitical environments to those in Central Asia, a situation that would lead to very different historiographies of East-West contact. Over the course of the nineteenth century, Egyptology and Assyriology both formed as distinct fields of archaeology, epigraphy, and philology. Together they identified traditions of writing stretching back three thousand years, cementing regions in Iraq and Syria as cradles of civilization. In Iran, the translation of a trilingual text offered testimony to the rule of Darius the Great.[74] By the late nineteenth century, much of the research conducted in the Near East sought evidence of the regional connections of antiquity. In the 1870s and 1880s, Heinrich Schliemann famously excavated prehistoric sites in Greece and Turkey in an attempt to confirm Homeric legends. As elsewhere, and as Swenson and Mandler illustrate, the archaeology of this period invariably advanced the imperial and territorial ambitions of European colonial powers.[75] Benjamin Porter indicates how such entanglements led to an intensity of effort in the coastal ports of Jaffa and Beirut, as well as in and around the strategically important cities of Aleppo, Damascus, and Jerusalem.[76] Research institutes provided a valuable excuse for establishing a presence in areas where European rivals anticipated diminished Ottoman rule, and large-scale excavations became an important source of military intelligence in the decades leading up to World War I.[77] This interface of military, political objectives, and archaeology was nowhere more apparent than in the work of Britain's Palestine Exploration Fund. The organization provided a home for the development of biblical archaeology, initially through the work of Sir William Matthew Flinders Petrie, George Reisner, and others. In Syria, excavations were undertaken throughout the second half of the nineteenth century by French, American, and English researchers. The establishment of the mandates system in the aftermath of the Great War further opened up Iraq, Lebanon, Palestine, and Syria to Western scholarship. Major advances in biblical archaeology during this period emanated from leading American universities, most notably via the University of Chicago's James Henry Breasted and William Foxwell Albright at John Hopkins University.[78] With the Old and New Testaments providing both the arc of time and geographical orientations of Holy Land scholarship, histories of Islam or connections across Asia outside India received little attention. The French Mandate in Syria also led to a distinct upswing in archaeological missions and scientific publications. Not surprisingly, Palmyra received extensive attention, with Danish and French teams excavating tombs and other parts of the complex in an attempt to understand its historical significance. The site had been known to Western scholars from accounts of European travelers to the region in the late seventeenth century. In search of the lands described by Homer, the antiquarian Robert Wood made an excursion to the site in 1751, documenting its architectural and artistic features. Wood's *The Ruins of Palmyra,* published in English and French, had a major impact on the design of neoclassical architecture in Europe and North America.[79] Palmyra also captured the imagination of European artists in search of the quintessential romanticism of an Orient in ruins, a highly aestheticized depiction that allegorically connected the site with the temples and monumental civic architecture of Rome and Greece. By the 1930s, the then-standard narrative on Palmyra's history reproduced in print and film related its rise and fall to Egypt, an imperious Rome, and the lives of Mark Antony and Cleopatra. A 1938 documentary made by World Window Productions and distributed by United Artists was one such example, with cinema audiences told in conclusion that "earthquake and successive plunderings by the Turk and Arab completed the destruction. The glory that was Solomon's now sleeps forever."[80] It is a narrative arc that remains prevalent, as illustrated in the statements of significance produced for Palmyra's listing as a World Heritage Site in 1980, a text that was updated in 2010:

> Recognition of the splendour of the ruins of Palmyra by travellers in the 17th and 18th centuries contributed greatly to the subsequent revival of classical architectural styles and urban design in the West. . . . The grand monumental colonnaded street, open in the centre with covered side passages, and subsidiary cross streets of similar design together with the major public buildings, form an outstanding illustration of architecture and urban layout at the peak of Rome's expansion in and engagement with the East. The great temple of Ba'al is considered one of the most important religious buildings of the 1st century AD in the East and of unique design.[81]

Palmyra thus speaks to the broader trends of early twentieth-century historiography on the Levant, which primarily framed the past through Hellenistic and Roman connections and the accounts of Alexander the Great, Herodotus, and Ptolemy, as well as the writings of more recent European travelers to the region.[82] The construction of this transregional history did more than just privilege the cultural and military achievements of European civilization; it played a critical role in conjoining modern Europe and the "Near East" in ways that served the ideological and territorial goals of European colonialism. Silberman has described this situation in the following way: "The modern discovery of the ancient splendours of the Near East by trained European and American scholars seemed to put Europe and America—rather than modern Egyptians, Palestinians, Turks, or Iraqis—in the position of legitimate heirs."[83]

By implication, there was much less interest in building a trans-Asiatic framework of East-West historical connectivity, despite knowledge about the travels of Marco Polo, expeditions to India by Alexander the Great, or the networks of silk trading between Rome and China.[84] This situation is emblematic of what John Hobson identifies as the

"Greek clause" in European historiography, or the widespread belief that "the Ancient Greeks were the original fount of modern (i.e., Western) civilisation."[85] In the teleological account of Western modernity and pioneering industrialization, he argues, East and West were divided into separate entities, with little or no recognition made of the scientific and cultural flows between them. Interestingly, Peter Frankopan opens his broadly conceived account *Silk Roads: A History of the World* with the same observation.[86] In 1934, the French archaeologist R. Pfister published the first of three volumes on textiles retrieved from the tombs of Palmyra. Among the fragments of cotton and wool were pieces of silk. Citing their geometric and animal designs, Pfister argued that the silk came from China. Subsequent volumes, published in 1937 and 1940, included chapters refuting attacks by those skeptical of his theory.[87] Clearly, analyses of transregional connections such as those produced by Pfister were marginalized by the prevailing theories and paradigms of the time, and it would take several decades for it to become commonplace to describe Aleppo, Baghdad, Damascus, or Palmyra as trading cities of the "famed Silk Road."[88] The discussion that follows argues that important seeds for such transitions would be sown in other contexts, such as travel writing, in the years leading up to World War II.

Traveling Eurasia

Tourism and accounts of adventurous travel have been key ingredients in the popularization of the Silk Road. The story, however, is not a simple one. The intention here is to briefly trace a number of geocultural imaginaries that came into existence through forms of long-distance travel across Eurasia, imaginaries that reflected the prevailing political conditions of their time but also continued to resonate through the twentieth century. As the idea of the Silk Road begins to surface in the run-up to World War II, we see how the popularization of figures like Marco Polo in the West play a role in "extending" the story into the Middle East. To begin, however, we return to the expanding road, railway, and steamship networks of the late nineteenth century. Nile Green has argued that the introduction of these infrastructures into Central Asia, together with the telegraph, heralded new ways of representing the region to European audiences. Accordingly, in introducing a volume on travel writing from this period he notes: "By the late 1890s, the Trans-Caspian Railway was already becoming a tool of culture by offering easier access to the region to artists, intellectuals, and journalists from Western Europe."[89]

Looking beyond the experience of European travelers, he also notes that this new age of mobility offered opportunities to aspirant tourists from within the region. The challenges they faced of passing through the frontiers of empire served as a precursor to the complications of the modern visa system, but stories of trips to neighboring countries or to Europe carried the same sense of romantic adventure as their European counterparts.[90] The recent study by Eileen Kane on "the Russian hajj" adds a further intriguing dimension to the history of intraregional mobility. Kane traces the routes and travel infrastructure established for hajj pilgrims across the Russian Empire in the nineteenth century.[91] The hajj provided vital, year-round revenue for Russia's rail lines and steamships, and to capture this income, special return tickets were created for pilgrims crossing the Caspian Sea and boarding trains along the Trans-Caspian Railway. As a consequence, those journeying from Central Asia were carefully steered away from more direct routes passing through Iran and instead kept within the borders of the Russian Empire by traveling through Baku or even via a much longer route that took them up through Oldenburg, past Moscow, and back down across the Black Sea.[92]

For European travelers, the railways were also central to the recommended itineraries for traversing Eurasia laid out in the first *Baedeker Handbook for Travellers on Russia, with Teheran, Port Arthur and Peking*.[93] Published in 1914, the travel guide included the now-familiar sections on the history, culture, and types of landscape that travelers were likely to encounter en route, as well as the logistical details of hotels, restaurants, and rail connections. While photographs of the antiquities of Central Asia were attracting attention across metropolitan Europe, the insights gleaned from the manuscripts and artifacts found in Dunhuang, Turfan, or Kashgar had yet to make their way across to the genre of travel writing.[94] It would be another five decades or so before the first guidebooks enticed readers along the Silk Road.[95] Instead, in the opening decades of the twentieth century, travel writing on Central Asia invariably focused on the sense of risk and adventure, and of crossing remote frontiers. The much-celebrated 1907 Peking-Paris road race opened up the possibility of travel by car. Essentially invented as a European media event, the race involved five cars setting off, each with a journalist, enabling stories to be telegraphed onward to editors in London and Paris. The makeshift infrastructure created along the route involved camels carrying gasoline to temporary service stations.

After traversing the Gobi Desert, the cars drove across Siberia toward Moscow, and on to Paris. Such events led to an increase in independent travel, including lone women travelers. In the 1920s, air travel opened up previously remote areas in Central and West Asia. However, by that point the tourism industries of large parts of the region had come under the purview of Soviet central planning. The establishment of Intourist as the official state travel agency of the Soviet Union in the late 1920s created a new geocultural imagining of Eurasia for Western tourists. Marketing campaigns for the agency's offices in Europe and the United States promoted road trips to the mountains of Georgia to see the "peoples of the Caucasus" or the landscapes of Soviet music.[96] Posters and brochures for the Trans-Siberian Railway promised a unique twelve-day experience of the diverse landscapes and cultures of a Soviet Union stretching from Europe to the "Far East" (fig. 2.5). And for those tourists wishing to travel farther south, tours by train visited "the modern comforts of Baku" and onward to the oil-producing regions of Iran.[97] In other words, within the Soviet Union, narratives of tourism were constructed to reinforce the idea of a union of integrated yet culturally distinct republics, a paradigm that had little interest in commemorating deep histories of cultural transmission and intra-religious dialogue.

By the time Hedin published his 1936 travelogue on the Silk Road, opportunities for excursions to Central Asia and China were coming to a close, even for his wealthiest of readers. In East Asia, World War II had already begun with the Second Sino-Japanese War breaking out in 1937, and Soviet Central Asia entered a particularly dark time with Stalinist purges, forced collectivization, and brutal labor camps affecting millions. Nevertheless, Hedin's writings found an audience at that time in large part because of the insatiable appetite in Europe and North America for adventure and remote travel, with exploits consumed widely in newspaper and book form, and increasingly in silent film. Around this time, the story of the archetypal figure of transcontinental travel, Marco Polo, was reaching new audiences. In 1938 Goldwyn released *The Adventures of Marco Polo,* a film that Iannucci and Tulk have argued dispensed with any sense of geographical specificity to present East and West through dichotomies of "civilized and primitive, active and passive, noble and wicked, male and female, dominant and submissive."[98] In this respect, the film sought connection with its audience through what are now the well-established tropes of Orientalism. But as Suzanne Akbari reminds us, the question is, which Orient?

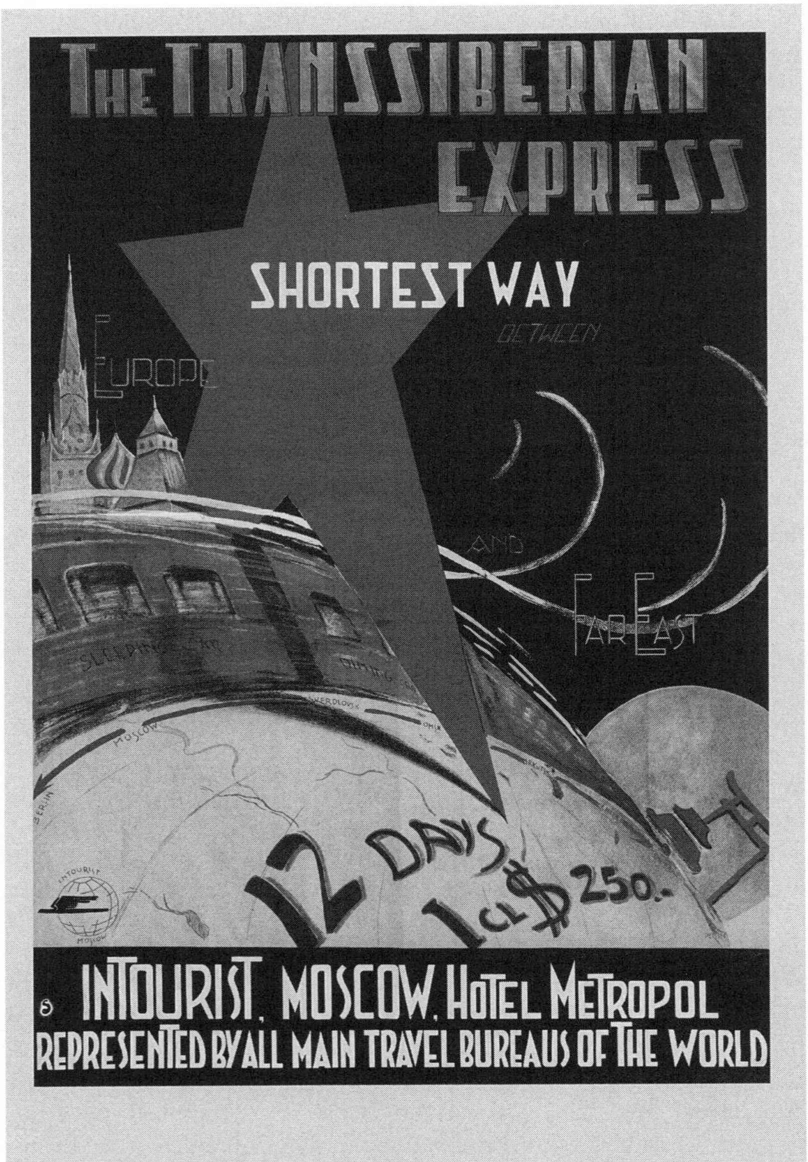

2.5 Trans-Siberian railway poster. Source: www.antikbar.co.uk.

Is it the erotic fantasies of Ottoman and Persian lands, the ascetic Buddhism of the Indian subcontinent, or the mysterious civilizations of Cathay?[99] The story of Marco Polo and its modern depictions embraced them all in a seamless geography of romantic adventure. For audiences of the 1930s, his story, reproduced on paper, film, and stage, connected the cities of the Holy Land with lands much farther east. In other words, in this quintessential story of East-West encounters, the Orient started in the Near East and stretched all the way through to Mongolia and China. The popularization of Marco Polo in this prewar period, infused with the locations and heroic figures of Middle Eastern antiquity gathered together from the writings of Ptolemy and modern scholarship extended the story of the Silk Road as a modern geocultural imaginary outward from Central Asia. A notable example of this is Lawrence and Margaret Thaw's account of a road trip across Asia. An article in *National Geographic Magazine* titled "Along the Old Silk Routes: A Motor Caravan with Air-Conditioned Trailer Retraces Ancient Roads from Paris across Europe and Half of Asia to Delhi" described their adventure from London to India, via Iran and the Holy Land. It opened in the following manner:

> The Great Silk Route! What visions of mile-long caravans of camels laden with spices and of the tramping march of invading hordes these words conjure up! Stretching east from Beyrouth or Antioch (Antakya) on the Mediterranean, it was known to Darius before Alexander and to the Assyrians before Darius. Imperial Rome used it as a direct means of communication with the East. By it Greek merchants, coming through Antioch, crossed the deserts of Mesopotamia (now Iraq). They paused at the mighty city of Baghdad before passing through the defile of the Zagros Mountains to reach the great trading center of Tehran and its near-by Caspian ports. After putting behind them the deserts of eastern Persia and the two-mile high passes of the Hindu Kush, they crossed the Khyber and other passes to the gold and spices of India.
>
> In the 7th century parts of this mighty caravan route of the dead past were traversed by the Chinese pilgrim Hsüan Tsang. Six centuries later Marco Polo followed similar parts of its tortuous course. Changes were made in the great land route from time to time as a result of new geographical discoveries or political complications, but the general direction remained the same.[100]

In the decades leading up to World War II, for those in Europe wishing to experience firsthand the adventures of the "exotic Orient," travel remained arduous and expensive. But as we have seen, quite different imaginings of East were in play, and certain routes were more accessible to tourists than others. The political events of the 1920s and 1930s that constituted the eastward expansion of the Soviet Union, together with the arrival of Intourist, put on hold the emergence of a Silk Road tourism industry across the Caucasus and Central Asia, a situation that began to change only in the 1990s. But the Thaws' expedition indicates how the geographies of the Silk Road were beginning to extend toward the Middle East through the writings of journalists and through accounts of the "Oriental East" depicted in film, television, and fiction.

East-West Encounters in the Shadow of the Cold War

In the aftermath of World War II, discourses of East and West took on a whole new complexion as the architecture of international affairs was remodeled in accordance with Cold War alignments. Significantly, here, this meant that a series of barriers, formal and informal, were erected, which together precluded the types of flows, mobilities, and connections described earlier. The idea of a historical Eurasian land bridge had little international currency in an era when East and West were defined by the ideological schisms of communism and capitalism and the map had been redrawn with decolonization. This reconfiguration of international affairs also marked a significant departure from the picture painted by Mackinder and his "pivot" at the beginning of the twentieth century, with Central Asia fading into the background for political analysts in the West.

In Asia, decolonization amplified the politics of history and heritage. For newly sovereign states, the project of nation building involved crafting narratives of the past in ways that shored up ethnic, cultural, and geographical borders. National icons and traditions needed to be symbolically enshrined, and, where necessary, invented. In Iraq, Pakistan, and Syria, archaeology was reoriented toward Islamic nationalism. Across the Soviet Union, researchers continued to use Marxist formulations of development and economic stages to interpret the histories of nomadic culture, emphasizing periods of static farming and group relations.[101] Elsewhere, Hinduism or Buddhism served as the bedrock for claims of homeland.[102] In other words, as political leaders across the region invested in the practices of statecraft, there was little impetus to nurture an identity politics rooted in histories of transboundary connectivity. This was particularly the case in Northwest China, where Mao worked to isolate Xinjiang from the rest of the country and use the region as a buffer to the Soviet Union farther west. As Gladney, Clarke, and others have documented, a strategy to achieve the latter thus involved the removal of "external influences through the neutralization of the region's historical, ethnic, cultural, religious and economic linkages to Central Asia."[103] Chin also suggests that the *Silk Road* term was rarely used in China at this time, indicating that it appeared primarily in newspaper articles on road construction in Xinjiang.[104] There were, however, a number of individuals and institutions that remained committed to putting this history back on the table, and it is here that we return to Japan.

The Bandung Conference, held in Indonesia in 1955, gathered together "the largest ever grouping of new entrants into the international system" and provided a platform for those nonaligned countries from Africa and Asia to express their frustration in the United Nations, and the destructive geopolitical environment being created by the Cold War.[105] The following year, a number of Asian leaders met in Tokyo to request that UNESCO take action to improve East-West relations in ways that better represented their interests and enabled a greater level of "mutual appreciation" between Europe and Asia.[106] In response, UNESCO launched the East-West Major Project on the Mutual Appreciation of Eastern and Western Cultural Values in 1957 to stimulate cultural relations and an interchange of "cultural values." The initiative ran through to 1965 and represented, as Wong puts it, "an unprecedented intergovernmental effort to engage states in dialogue around cultural identities in the midst of redefinition and rising ambiguity about the meaning of East and West."[107] Dozens of conferences, exchanges, publications, and international collaborations addressed questions of cultural transmission and exchange through the lenses of folklore, music, craft, archaeology, religion, and art.[108] In the 1950s and 1960s, China was yet to develop a substantive relationship with UNESCO. The victory of the Chinese Communist Party in the country's civil war in 1949—which led to the creation of the People's Republic of China and the retreat of the Nationalists to Taiwan—initiated a period of complex diplomacy with the United Nations and its other founding members: the United States, Soviet Union, and United Kingdom. With the United States and its allies recognizing the Republic of China as the sole legitimate government of China until 1971, much of the activity from Asia in the East-West Major Project came from India and, to a lesser extent, Japan.

A few months after the project's launch, the Japanese National Commission for UNESCO organized the International Symposium on the

History of Eastern and Western Cultural Contacts.[109] As part of the preparations, around twenty Japanese specialists produced a background document, *Research in Japan in History of Eastern and Western Cultural Contacts: Its Development and Present Situation*.[110] As the title suggested, the report summarized the history of Japanese scholarship on Asia over the course of the twentieth century, incorporating the work of Ōtani and his team. But with the geographies of East-West contact once again seen from a Japanese perspective, the focus primarily fell on understanding intra-Asian connections, with Turkey often constituting the "far west." The publication made a number of references to the Silk Road, but it also made clear that the existing conceptualization of a land bridge between China and the Mediterranean was too narrow for understanding the depth and complexities of a connected Asia.[111] To consolidate the different strands of research undertaken by Japanese scholars, the report identified three routes: Oasis, Steppe, and Sea. An introductory essay by Hisao Matsuda recognized the strong influence of European scholarship for understanding the historical ties along the Oasis Route but argued that, for all three routes, Japanese researchers made important contributions to knowledge. Accordingly, he suggested that "the discovery of the Steppe Route best reveals the creativity of the Japanese orientalists."[112] To justify this, Matsuda pointed to the work on Chinese relations with Mongolia and India, as well accounts of ninth-century Uyghur migration, to argue that a focus purely on east-west flows misses histories of north-south connectivity and important changes within Central Asia. Japanese studies of tribes in Mongolia in the steppes of northern Asia rectified this and challenged conventional ideas about nomads as the inferior barbarians of a more civilized Chinese.[113] Finally, the report laid out early twentieth-century Japanese scholarship on maritime connections in East and Southeast Asia. The focus was on trade connections established by Japan and China in the centuries before European ships arrived in the area. It is here that we see the idea of a sea route, or the Maritime Silk Road, articulated for the first time in an international policy setting.

This initiative, together with the larger umbrella project led by UNESCO, provided Japan with a platform of international cooperation for rebuilding diplomatic relations with the West and with its regional neighbors. Through the Silk Road, those on the political left in Japan were able to project the image of the country as a civilization whose culture and religion had been in peaceful dialogue with others over many centuries. To this end, an important opportunity arrived in 1972 with the historic visits to China by Japanese prime minister Kakuei Tanaka and US president Richard Nixon. During Tanaka's visit, a director from Japan's state broadcaster, NHK, proposed a documentary on the Silk Road as an initiative for rebuilding Sino-Japanese relations. After years of negotiation, filming started in 1979 as a partnership between NHK and China Central Television, with Japanese film crews granted unprecedented access to China's northwestern provinces at a time when the wider region was entering the turmoil of the Soviet-Afghan War and Iran-Iraq conflict. Shown over twelve parts in Japan in 1980, the series covered the region from Xi'an to the borders with Pakistan and the Soviet Union. With the addition of an English soundtrack, it was subsequently broadcast in more than thirty-five countries to critical acclaim. Around this same period, NHK was also involved in producing a second television series set in Northwest China, but one that was altogether different from the Silk Road documentary. *Saiyūki* was broadcast over fifty-two episodes in Japan between 1978 and 1980. Dubbed and renamed *Monkey* for international export, the series retold one of the most famous works of Chinese literature, *Journey to the West*, which has been attributed to the sixteenth-century author Wu Cheng'en. The novel depicts the travels of the seventh-century Buddhist monk and scholar Xuanzang to Samarkand, Balkh, and down into India. Fantasy action scenes and a memorable soundtrack helped *Monkey* achieve a cult following in multiple countries.

Japanese interest in the Silk Road continued to grow through the 1980s, and in 1988 the city of Nara collaborated with NHK to host *The Grand Exhibition of Silk Road Civilizations*. Within Japan, Nara is commonly referred to as the final stop on the Silk Road, and the $83 million exhibition triangulated the city's historical significance by again identifying three Silk Road route themes. In a modification to the study of 1957, the two land routes were combined into the exhibition *The Oasis and Steppe Routes*, with the other two exhibitions titled *The Route of Buddhist Art* and *The Sea Route*. Collaborations were established for each, with a number of countries importing artifacts for display. For *The Oasis and Steppe Routes* the key partners were government ministries from Italy, Iraq, the Soviet Union, and France. Each sent items, including the Musée Guimet in Paris, which supplied artifacts from Afghanistan collected by Paul Pelliot and others. Japanese museums supplied items from Iran. For *The Route of Buddhist Art*, items were procured though formal arrangements with India, China, Korea, and Pakistan. Finally, for *The Sea Route*, artifacts were sourced from the Syrian Ministry of Culture.

Three elaborate exhibition catalogs were produced for the event, each of which contained essays about Japan's place in a long history of cultural transmission and regional trade. To complement the exhibition's objects, Namio Egami offered a narrative of the Silk Road built around the accounts of long-distance travelers from within the region. In an effort to counterbalance the east-west arc of Marco Polo, a number of Asian travelers—namely Faxian, a Chinese Buddhist monk; Zhufahu, of Kushan descent and born in Dunhuang; Changhun Zhenren, a teacher of Taoism; and Xuanzang—were cited to tell the story of religious, cultural, and technological transmission across the region.[114] The catalogs also provided an opportunity for ambassadors and ministers of culture involved in the exhibition to express their thoughts and reasons for participating:

Pakistan is happy to contribute to this Exposition by sending a special exhibition of Gandhara sculptures depicting the culture of the people who lived here and followed Buddhism during the early centuries of the Christian era. Gandhara art is itself a manifestation of the cultural traditions of Greco-Roman art. . . . I hope that the exhibition will provide a stimulus for the propagation of world peace through these master pieces of art and culture.
NISAR MOHAMMAD KHAN, MINISTER FOR CULTURE, SPORTS, AND TOURISM, GOVERNMENT OF ISLAMIC REPUBLIC OF PAKISTAN

It is of great significance that Japan is hosting the Silk Road Exposition in Nara concurrently with the Seoul Olympics, to take place in September by virtue of cooperation between East and West. The Silk Road, historically linking East and West in cultural exchange, served an essential pathway of cultural empathy. . . . I am sure that this Exhibition will further the mutual understanding between East and West.
CHUNG HAN MO, MINISTER OF CULTURE AND INFORMATION, REPUBLIC OF KOREA

Today as science and technology continue to progress rapidly, a need is strongly felt among us to reexamine our past and to try to understand the principles of cultural development with reference to the origins of civilization. As we look into the histories of peoples around the world, we realise that communications beyond national boundaries, for instance, exchange of ideas and discoveries, cultural exchange and trade, have always been major factors contributing to social development. Needless to say, one outstanding symbol of such meaningful, rich exchange among various peoples and cultures of the world is the Silk Road, a great link between East and West. . . . I would like to express my sincere gratitude and respect for your inter-

est in our exhibits and my strongest hope that this exhibition will be of some help towards mutual understanding between the peoples of Japan and the Soviet Union.
GURII IVANOVICH MARCHUK, PRESIDENT, ACADEMY OF SOCIAL SCIENCES, SOVIET UNION

Syria has been one of the most fortunate beneficiaries of civilization, making and assimilating human heritage, that ever ambitious force of creation and creativity that has left us testimony after testimony to the grandeur Syria was and is—an uninterrupted line of creative development encompassing past achievements as well as contemporary technological prowess. . . . Japanese archaeologists and foundations have played a major role in the search for treasures of Syria's past, both on land and offshore. . . . And it is in recognition and appreciation of this fruitful cooperation with our Japanese colleagues that Syria has decided to participate in The Grand Exhibition of the Silk Road Civilizations in coordination with the Nara Authority and NHK. Such forms of cultural cooperation, we are confident, are bound to yield fruit in deepening understanding of and knowledge about Syria whose present seeks to match its past with tireless efforts by its president, Hafez Assad, to encourage not only digs into the past but also inroads to the future so that Syria's contribution today may equal, or at least not fall short of, its contribution to the civilization of yesterday.
NAJAH ATTAR, MINISTER OF CULTURE, SYRIAN ARAB REPUBLIC[115]

Two decades after UNESCO's East-West Major Project finished, its guiding principles of mutual appreciation and civilizational dialogue are clearly evident. But in this discourse of international cooperation, we also see a number of shifts in the appropriation of the expanding geographies of the Silk Road. By this point it had become a clear instrument for Japanese peace diplomacy at the bilateral and multilateral levels, and it is emerging as a space of intergovernmental cooperation between non-Western countries, an antecedent to today's South-South cooperation. Interestingly, the exhibition led to long-term archaeology and heritage management aid from research institutes in Nara for Palmyra, with projects ongoing today.[116] Finally, we also see the early signs of the Silk Roads being deployed as a platform on which international development aid and ideas of progress are folded in with antiquity and conjoined pasts. In other words, the exhibition provided an expansive discursive template for cooperation, which remains present in Belt and Road today.

In the aftermath of World War II, China and the Central Asian states of the Soviet Union were effectively closed to Western tourists and journalists. Luce Boulnois was a rare exception. Gaining access to remote areas through her role as a translator and researcher at France's Centre national de la recherche scientifique, Boulnois published one of the few books on the history of Central Asia, including China, of the 1960s. Her 1963 *La route de la soie,* translated and republished as *The Silk Road* three years later, focused primarily on the production and trade of silk itself.[117] In this regard, the book contributed to the reification of silk in the popular imagination of the road, contributing to its allure and romantic appeal. Interestingly, though, her geographies of east-west encounters were once again fashioned around stories of Marco Polo and the geopolitics of transboundary infrastructure. The international success of the NHK series some years later occurred at a time when Central Asia had been thrust back into the spotlight through the Soviet invasion of Afghanistan. Expert media commentary on these events also gave new life to the Great Game metaphor. But here we also see some interesting turns in how journalists and travel writers depicted the geography of the Silk Road. In researching his book *An Adventure on the Old Silk Road,* the BBC journalist John Pilkington was forced to skirt along the political frontiers of Cold War Central Asia. With much of the East-West route depicted by Richthofen off limits to him, Pilkington once again "retraced" the route of Marco Polo through Turkey, Pakistan, and up along the Afghan border to Kashgar. In a similar vein, Jan Myrdal's 1980 travelogue, *The Silk Road: A Journey from the High Pamirs and Ili through Sinkiang and Kansu,* recounted the frustrations of obtaining travel permits for Afghanistan and a lifelong ambition of completing "a real journey: following Marco Polo's route eastward."[118] Myrdal's account of the trip focused heavily on the role of Xinjiang in twentieth-century Sino-Russian relations and the ways in which the region was being transformed through master plans targeting industrialization and modernization.

Looking beyond the Japanese examples noted earlier, international interest in Central Asia and the themes of a Silk Road history continued to grow in the 1980s. To cite a few examples, an American-Italian collaboration led to the internationally acclaimed television miniseries *Marco Polo.* Silk Road museum exhibitions began to appear in different cities, with *Silk Roads: China Ships,* held in Toronto in 1983, notable for its presentation of maritime trade and cultural transmission.[119] Within China itself, the 1980s period also saw growing awareness of the country's cultural connections with Inner Asia. A twenty-five-part television adaptation of Wu Cheng'en's *Journey to the West* became an immediate success in the late 1980s, as did Yu Quiyu's travelogue explication of Chinese culture and history, republished in 1992 under the title *A Bittersweet Journey through Culture*. In a chapter on Xinjiang, he detailed the rich mix of cultural influences and civilizations brought along the Silk Road, describing Kashgar in the following terms:

> This was a place where all travelers, explorers, Buddhists, and merchants had to stop. No matter going out or in, they all suffered this harsh testing ground, but they faced greater tests before, maybe the Pamir, maybe the Takla Makan Desert. Therefore, they purged their souls, picked themselves up, and prepared to risk their lives on the road again. For many people, this was the last station for their lives; for others, this was the new starting point, filled with generousness, fortifying them for their subsequent departure. Whether as terminal or starting point, here was a place for heroes' pouring wine and sacrificing. Every inch of air in Kashgar bore the dumb guttural sounds of such men. The world yearned for the connection from here time and time again.[120]

The following chapter also indicates how the figure of Zheng He was beginning to enter the lexicon of Chinese politics and diplomacy at this time. Of greatest significance, though, was the internationalization of China during this period, together with a reduction in Cold War tensions and the subsequent collapse of the Soviet Union, both of which proved catalysts in raising the international profile of the overland Silk Road. Previously inaccessible areas and remote historical cities began to receive elite tour groups from Europe and North America, with itineraries combining camel tours with private jets and heritage hotels. Sweeping political changes also provided new opportunities for international collaborations in the cultural sector. In 1989, the Los Angeles–based Getty Conservation Institute initiated a project for heritage conservation at the Mogao Caves outside Dunhuang. Five years later the International Dunhuang Project launched as a collaboration among institutions in London, Beijing, Dunhuang, Berlin, Kyoto, and St. Petersburg. Managed by the British Library, the project continues to conserve and catalog those manuscripts and artworks that traveled out from Dunhuang at the beginning of the twentieth century.[121] With financial support from Oman, Japan, Korea, and France, in 1988 UNESCO revived its ambitions of fostering East-West cooperation through the Silk Roads: Roads of Dialogue Project, an initiative that ran until 1997.[122] As UNESCO's director general Federico Mayor indicated, the Silk Roads powerfully embodied the organization's mandate for advancing international dialogue at a time of significant instability and change:

Through this project, UNESCO has sought to shed light on the common heritage, both material and spiritual, that links the peoples of Eurasia. To generate an awareness of the different civilizations' shared roots and to foster the concept of a plural world heritage that embraces the masterpieces of nature and culture in all countries is, in the final analysis, to encourage attitudes of openness and tolerance, so necessary in an essentially interdependent world. The fundamental issue at stake in the "roads of culture" approach is to highlight the significance of pluralism in culture, no less vital than that of biodiversity in nature.[123]

More than twenty-five international conferences and colloquiums led to a series of publications that further diversified the geographies and themes of the Silk Roads. Once again, three routes—oasis, steppe, and sea—oriented the project.[124] But by drawing on new research, these events further expanded the framing by addressing the topics of nomadic culture, harbor cities, food, and Arab seafaring, as well as less familiar routes of connectivity, including the regional and long-distance connections of East Africa and Southeast Asia.[125] A decade of events and publications had a significant impact on raising the visibility of the Silk Roads across agencies engaged with international cultural policy, and, in so doing, gave attention to a geographically and thematically expansive definition of Eurasian connectivity. By the mid-1990s the idea of the Maritime Silk Road also began to gain visibility among cultural institutions in southern China. A collaboration between the Guangzhou Museum and Hong Kong's Urban Council led to a large exhibition titled *The Maritime Silk Route: 2000 Years of Trade on the South China Sea*.[126] A few years later, the Smithsonian Institute in Washington, DC, took up the theme via *The Silk Road: Connecting Cultures, Creating Trust* festival. Located on the Washington Mall, their annual two-week festival has become a unique forum for cultural diplomacy. The 2002 event featured Yo-Yo Ma's Silk Road Ensemble, formed two years previous.[127] In his capacity as festival curator, Richard Kennedy noted the significance of its timing:

The idea of the Silk Road is still available for new interpretations. And in the present political environment the idea is particularly evocative. One reason Smithsonian staff has been particularly excited to work on a Silk Road project at this time is the political transformations that have taken place in the region over the previous two decades. The opening of China and the collapse of the Soviet Union have enabled researchers, businessmen, and travelers alike to visit a vast area little known to Westerners in the past hundred years. A new Silk Road is being traveled. The modest victories of democracy and capitalism at the end of the second millennium allowed strangers once again to meet along the ancient roads of silk and once again exchange ideas and products. People spoke of new economic and political realities, and it seemed that new cultural realities were likely developing out of this transformation as well.[128]

The UNESCO and Smithsonian initiatives reflected wider shifts in scholarship occurring at the time. In what seemed to be an increasingly borderless world of 1990s globalization, conceptions of space and capital were recast around mobilities, networks, and flows. Flood and Necigpoğlu note that their own field of Islamic art history expanded outward from the Middle East "to encompass regions and periods traditionally excluded from the canon."[129] This established connections with earlier work on world and transregional history, as well as theories posed in the 1980s on trade networks across the Indian Ocean, a theme picked up in chapter 6. Wider changes of the 1990s also led to a renewed interest in the concept of civilization, albeit in very different ways in different disciplines. The end of the Cold War had triggered a shift in the Western commentary on East-West relations. Civilization entered debates in political science and international affairs, most notably through the ideas of Samuel Huntington and his followers. His "clash" thesis framed by the fault lines of culture and religion found its reaffirmation in 9/11. Hannerz traces the international circulation of such arguments and their refutation to argue that a widely held feeling that the world was entering a new era of complexity and shifting borders gave unprecedented prominence to culture in the analysis of geopolitics.[130] Central Asia had once again returned to being a critical "pivot" in world affairs, but in contrast to Mackinder's early twentieth-century thesis, this time the crucial issue was the security of transcontinental energy supplies. Increasing demand for oil and gas from China and India, together with Middle Eastern conflict and the presence of terrorist groups such as Al-Qaeda in Afghanistan and Pakistan, put new demands on the global energy sector. The clash-of-civilizations model thus looked to the inner zones of Asia as an area of paramount importance to global security.

But interestingly, the landscapes of the Silk Road also came into focus for those invested in its history as a "crossroads of civilization." In sharp contrast to the ideas of Huntington, others responded to late twentieth-century globalism by questioning the practice of framing civilizations and cultures as bounded, self-contained entities. The turn toward networks and mobilities facilitated the cultural theorization of cosmopolitanism and hybridity. This gave the story of a Silk

Road cosmopolitan culture new resonance, a standing that was further enhanced by a flurry of international conferences and through fresh discoveries and interpretations. In Palmyra, for example, the excavation of the first silk fragments since the 1930s was correlated with recent evidence of looms and other weaving technologies in China dating back to the Han dynasty.[131] New ideas were also filtering through to Silk Road documentary making. Shortly after the UNESCO project finished, production started on the series *NHK World Heritage 100*. In their film on Palmyra, NHK departed from the story of Rome and Cleopatra, as presented by World Window Productions in 1938, in favor of merchants, and the city as a crossroads of civilization and Silk Road caravansary.[132] The chapters that follow offer a number of examples—China's increased investment in maritime archaeology institutes, the approbation of Zheng He, scholarly conferences dedicated to maritime Islam and Southeast Asia—to illustrate how this shift toward transnational and transregional understandings of the past has continued to gather momentum.

Conclusion

In 1997 the US Congress introduced the Silk Road Strategy Act.[133] Despite repeated attempts, the bill never passed. In broad terms, its aim was to strengthen democracy and foster economic development through enterprise and infrastructure projects in the South Caucasus and Central Asia. By 2003, and in the wake of 9/11 and the US invasion of Iraq, a number of foreign policy think tanks in Washington once again emphasized the importance of nonmilitary interventions in these two regions. An Obama administration looking for new ideas positively received a strategy for building infrastructure and trade capacities put forward by the Washington-based analyst S. Frederick Starr. Aware that securing financial commitment from within the United States was a politically complicated proposition, Starr pointed to the importance of local ownership. Eurasia, he suggested, had once again reached a critical juncture:

> Thanks to the collapse of the USSR, whose closed border stood like a wall across the heart of Eurasia, to China's decision to open trade across its western border, and to the gradual return of Afghanistan to the community of nations, continental trade spanning the entire Eurasian land mass is again becoming possible. Western Europe, China, the Middle East, and the Indian sub-continent can, in time, connect

with one another and with the lands between by means of direct roads, railroads, and technologies for transporting gas, oil, and hydroelectric power. These "new Silk Roads" have enormous potential for the entire Eurasian continent, and especially for the countries of "Greater Central Asia" which they must traverse.[134]

The previous chapter noted the development of various multilateral and unilateral initiatives for transboundary infrastructure during this period, and in 2011 Hillary Clinton followed these up with the launch of the New Silk Road Initiative. The project failed to achieve its goals, but the idea had arrived of a new Silk Road for the modern era, envisaged around pipelines, transcontinental rail, energy production, and cross-border trade. But what Starr and Clinton's strategy lacked was an understanding of how their new Silk Road mapped onto a narrative of history being propelled forward by rapid changes in the region. The Silk Road is a story of internationalism, and as this chapter has demonstrated, such historical narratives require a particular ecology in order to blossom. The arc of depicting collective pasts in the twentieth century was defined by the interplay of nationalism and internationalism. Across Asia, the former invariably remained the prevailing paradigm, particularly in the contexts of postcolonial nation building and the nationalist imperatives advanced in the creation of the Soviet Union. Late twentieth-century globalization complicated this balance, and the rise of the Silk Roads as a brand and concept on the international stage during this period speaks to this shift. But as this and subsequent chapters indicate, it is a narrative of transregionalism and internationalism that is continually subverted by national interests. In this regard, China follows the footsteps of India, Japan, Turkey, and others.

In tracing the biography of the Silk Road as a concept since the late nineteenth century, we see the different ways the Silk Road has proved a compelling landscape of the imagination. Tales of travel to far-off lands passed down from the accounts of late nineteenth-century explorers, and the enduring fascination of figures likes Marco Polo and Xuanzang embody a sense of adventure. For today's tourists the romance of the road comes from the promise of traveling back in time, meeting cultures "untouched" by modernity. And for the real romantic, adventure holidays retrace the routes of camel caravans or the footsteps of famed travelers. In addition, it has been argued that this grand narrative of connectivity has come to be imbued with an aura of cross-cultural exchange and cosmopolitanism, themes that have permeated the worlds of academia, cultural policy, film, and television, among others, as well as forms of international cooperation. But as we have also seen, once such ideas enter the museum and world of intergovernmental policy, or the speeches of politicians and diplomats, they are encased in a language of peace and international harmony. Chapter 5 takes up the question of how these themes are embodied in objects of antiquity and extends the discussion of Silk Road exhibitions into the new heritage diplomacy of Belt and Road.

In constructing a biography of the Silk Roads it is also evident that the concept is deeply entangled in international relations and imperial ambition. Silk Road discourses gain ascendancy when they can affect some political work. As the significance of Central Asia to world affairs has pivoted, so, too, has the significance of a history of transregional connections. The opening chapter illustrated why this now extends to the Maritime Silk Road vis-à-vis the growing significance of the Indian Ocean region. But in looking at the experiences of India and Japan, it also becomes apparent that as narratives of the Silk Roads constructed within Asia privilege intraregional connectivities, the concept takes on particular political expediencies. The early threads of pan-Asianism we see here were rooted in the ideologies of anticolonialism. This raises intriguing questions as to whether the latest iteration of the Silk Roads as formed through Belt and Road also carries undertones of pan-Asianism, one that is quietly exclusionary to those outside the region. Chapter 7 considers this issue further.

Finally, in addressing those other evocative metaphors associated with Silk Road regions—the Great Game and Cold War—the chapter has highlighted the deep entanglements of archaeology, material culture heritage, and the geopolitics of transboundary infrastructure. It is here, then, that we can begin to offer a more substantive response to the proposition that Belt and Road represents the revival of the Silk Roads for the twenty-first century. Moving beyond an answer that sees the historical context of camel caravans and sailing ships as mere romantic metaphor, a biography of the Silk Roads in the modern era suggests that China is reviving a theater of geopolitics and great-power accumulation, a vision of cross-cultural dialogue and civilizational grandeur, and an imagined landscape where the rewards of traveling over great distances are plenty. This, along with a parsing of the different ways in which antiquities are enmeshed in expansive nationalisms, the politics of cooperation, or discourses of regionalism rooted in culture and history helps us better grasp what's at stake in Belt and Road.

THREE

A Politics of Routes

The Restitution of Greek Culture

In 2010 Greece found itself in a storm of international media attention, with journalists and reporters extracting as much metaphorical blood as they could from the stones of Athenian antiquity. The government debt crisis, we were told, was a Greek tragedy of monumental proportions, one that threatened to undermine, if not shatter, the liberal order that had been built up over the previous two decades and would have ramifications that stretched far beyond contemporary economic and political affairs. Within the intense debates that raged across Europe and beyond concerning the possibilities of "exit" and "breakup," many wondered whether we were witnessing a crumbling of European civilization itself. It was a humiliating prospect; Greece and Athens were, after all, the seats of European culture. The strident positions held by Angela Merkel, François Hollande, and others centered on tensions between critically exposing and reforming poor Greek productivity and overspending on the one hand, and holding together a family of nations that shared a history and heritage stretching back millennia on the other. In the name of efficiency and debt alleviation, Greece was called on to privatize its key state assets. By early 2015, €3.2 billion had been raised, a small fraction of the €50 billion needed, with the Greek government refusing to conduct a "fire sale" of the country's key infrastructure assets. In June that year, they broke off negotiations with the European Central Bank, International Monetary Fund, and European Commission regarding the terms of the second bailout. Alexis Tsipras took the decision to hold a referendum a month later, with Greeks overwhelmingly voting to reject the punishing conditions of austerity being imposed upon them. Journalists across Europe were aghast at a country they perceived as having few other options than to accept the terms of aid and the need to undertake major reform. What they missed, however, was how the prospect of "Grexit" gave new weight to conversations and openings that drew Greece into a much larger orbit of geopolitics and global trade. For Russia and China, Greece's predicament represented significant opportunities, and for China, in particular, securing control over key infrastructure resources in the Mediterranean aligned perfectly with Xi Jinping's ambitions of building an integrated Eurasian economy.

It was within this context that a small but extraordinary conference took place at the University of Athens in September 2014. The event, "The New Silk Road: Cultural and Economic Diplomacy China-Greece-Europe," centered on the conjoining of Greece and China, both past and present. Professor Zhang Lihua, director of the Research Centre for China-EU Relations at Tsinghua University, opened the discussion by highlighting the historical bonds the two countries shared. They were not merely nation-states but civilization-states with diverse traditions. Such diversity, she suggested, in Greek philosophical and political thinking revealed that Greece was not merely European but Mediterranean and Asiatic as well. In both countries, philosophers provided the cultural and moral foundations for two great empires of Eurasian history—the Greco-Roman world and the Han-Tang Chinese world—whose influence was felt far and wide. To express empathy for Greece's humiliation and struggle for sovereignty, Zhang described how Western agencies and governments had condemned and coerced China over Taiwan and human rights issues since the end of World War II. As a parallel, Christodoulos Yiallourides, president of the Hellenic Foundation for Culture, suggested that the ties between "Greeks and Chinese [are] not merely political and economic but also metaphysical and deeply cultural."[1] For Yiallourides, diplomatic relations between the two countries can help Greece endure its problems and provide China with an entry point into Europe for its business interests and its art and culture. As part of a published summary of the event, Vasilis Trigkas stated that "Yiallourides expressed his worries on the current situation in the Middle East and on the need to reform the liberal world order, which was created by the winners of WWII. Democracy has been misinterpreted and used to promote the interests of a superpower. Real democracy is political and demands mass participation. He expressed his confidence that China's role in that immense global transformation has been positive and that a stable and strong China will benefit world harmony, peace and global problems."[2]

Subsequent presentations from economists, architects, sociologists, and historians reinforced the idea that the philosophical and cultural foundations underpinning the economic and political systems of both countries could help address the security and global inequality challenges of the twenty-first century. One proposal focused on developing Sino-Hellenic cooperation around monument preservation and research as a way to better understand these shared values. Significantly, such ideas were not confined merely to the corridors of academia. Three months earlier, in June 2014, on his first visit to Greece, the Chinese premier Li Kepiang signed agreements with then prime minister Antonis Samaras regarding cooperation across the fields of politics, trade, business, and culture. Plans were put in place for a Chinese cultural center in Athens dedicated to heritage preservation, underwater archaeology, and Olympic culture, with a sister center to be established in China.

Events over the following two years would reveal the larger strategic purpose of this celebration of shared civilizational values and culture. On a number of occasions, Zou Xiaoli, the Chinese ambassador to Greece, publicly expressed his government's commitment to building bilateral ties across multiple sectors. But it was the signing of a contract in early 2016 that would bring the various elements together. On April 8, Alexis Tsipras formalized an agreement with the China Ocean Shipping Company, or COSCO, for a 67 percent majority stake in the Piraeus Port Authority, near Athens. Immediately injecting nearly €400 million into the Greek economy, the deal included long-term funding commitments designed to make Piraeus the largest port terminal in the Mediterranean by 2020. Reports commissioned to assess the project estimated long-term revenue to be around €5 billion per year, with more than one hundred thousand jobs being created. To commemorate this new era of cooperation, a conference was held in Athens on Greece and the new Silk Road. Speaking to an audience of business leaders, politicians, and diplomats, the ambassador Zou Xiaoli stated:

> As the cradle of Western civilization, Greece, in its long course of history, made outstanding contributions to the cultural integration and mutual learning between the East and West through the ancient Silk Road. Today, the COSCO Piraeus project has once again placed Greece at an important knot of the Silk Road, a new Silk Road

formed by the "Silk Road Economic Belt" and "Maritime Silk Road of the 21st Century." It provides a once-in-a-thousand-year opportunity for China and Greece, the two ancient and great nations, to join hands in promoting the progress of human civilizations. I hope and believe that, with the relentless efforts of our two countries and two peoples, we will succeed in building the COSCO Piraeus project into a shining example of the "Belt and Road" Initiative.[3]

His speech also contained details of China's larger ambitions for the region. The intention was to connect up to a newly built Hungary-Serbia railway, the two projects constituting the European leg of the China-Europe Land-Sea Express Route. Piraeus, it was suggested, could thus be a trade catalyst for the entire region and a gateway project encouraging further Chinese investment across multiple sectors:

The China-Europe Land-Sea Express Route, with the Port of Piraeus as the hub, will bring down remarkably the cost of time, energy, and capital for the flow of commodities between China and Europe and between Asia and Europe. The Express Route will also link the new Maritime Silk Road with the Silk Road on land and strongly boost economic and trade cooperation among China, Russia, Central Asia, Central and Eastern Europe, Southeast Europe, West Asia and North Africa, and Southeast Asia.

As the COSCO project grows, China-Greece relations are also flourishing in the fields of tourism, culture, education, science and technology, press and media, and academic studies. . . . Cultural heritage preservation, underwater archaeology, the Marathon race, the Olympic culture and the cultural industry will be the key areas of cooperation for this year. . . . The day before yesterday, I had the special honour to be invited to speak at the signing ceremony of the cooperation protocol on the establishment of the "Amphictyony of Ancient Greek cities." I assured the Greek mayors of my commitment as the Chinese Ambassador to bridge the Greek cities with the Chinese cities in order that the peoples and local authorities can play a bigger role in the further expansion of our bilateral relations.[4]

The story of Greece sets the scene for the following two chapters. In these excerpts we see the prospect of present-day connectivity coupled with the idea of not just economic but also civilizational revival. Multibillion-euro port deals and the promise of new shipping routes are the driving force for cooperation across multiple sectors, energizing a shared enthusiasm for discovering cultural and even metaphysical affinities rooted in deep histories, and excavating the spiritual and material evidence of a rich exchange between civilizations. In this desire to reclaim such pasts, the fluidity and malleability that the Silk Road proffers appears once again. Much of the evidence of historical contact between the two countries points to the trade networks of the overland routes, and yet the excitement of seeing Athens develop as a key node in Belt and Road stems from its strategic location on the Maritime Silk Road of the twenty-first century.

It would be safe to assume that the Chinese Communist Party's original idea of reviving the "ancient" Silk Roads was metaphorical. As far as I can tell, no one has actually proposed to reestablish camel caravans for trade or markets dedicated to horses and slaves. The vast majority of reports from think tanks and media outlets published in the wake of Belt and Road's launch documented the transboundary trade and infrastructure routes it would create. To accompany these analyses, graphic designers produced a cartography of smooth, transcontinental lines, visualizing how the Silk Road would be rebuilt for the twenty-first century. In the flurry of excitement around Belt and Road, a normative history quickly settled in, one that draws on the Silk Road as a metaphor for trade, exchange, and connectivity. Such themes will remain central to the initiative, but few reports in the Western media and academia noticed the other metaphorical threads woven into the diplomatic discourse of Belt and Road as it evolved. In focusing on a series of speeches by politicians, diplomats, prime ministers, and presidents over the first three years of the project, my interest in the following pages is to explore the more subtle, but perhaps far more powerful, metaphors and ideas that have come to be embedded in the appropriation of a Silk Road history and heritage. The aim, then, is to explore what expediencies the idea of "reviving" the (ideas of) peace, harmony, friendship, trust, cooperation, spirit, and prosperity of the Silk Roads holds in the twenty-first century. Chapter 4 builds on this by addressing how such themes have come to be materially enacted along certain corridors via festivals, archaeological sites, expos and in policies for tourism and urban planning.

Connecting Futures: Metaphors of the Past

Xi Jinping's twenty-five-minute speech to Nazarbayev University in Astana in 2013 outlined a strategy for the regional integration of economies and infrastructure. He identified bodies such as the Eurasian Economic Community and the Shanghai Cooperation Organisation as key mechanisms for cooperation, signaling their role within a series of broad aims: promotion of unimpeded trade, better communication around policy, improved road connectivity, enhanced monetary circulation, and an increased understanding between people in the region. This initial draft of what would become the five pillars of Belt and Road was presented as a joint venture toward "a new economic belt along the Silk Road." The speech also featured an early draft of how the region's historic connections were to be reactivated. According to Xi Jinping, the project connected nostalgia for the geographies of his own life story and that of the ancient Silk Road:

> Over 2,100 years ago during China's Han Dynasty, a Chinese envoy Zhang Qian was sent to Central Asia twice with a mission of peace and friendship. His journeys opened the door to friendly contacts between China and Central Asian countries as well as the Silk Road linking east and west, Asia and Europe. Shaanxi, my home province, is right at the starting point of the ancient Silk Road. Today, as I stand here and look back at that episode of history, I could almost hear the camel bells echoing in the mountains and see the wisp of smoke rising from the desert. It has brought me close to the place I am visiting. Kazakhstan, sitting on the ancient Silk Road, has made an important contribution to the exchanges between the Eastern and Western civilizations and the interactions and cooperation between various nations and cultures. This land has borne witness to a steady stream of envoys, caravans, travelers, scholars and artisans travelling between the East and the West. The exchanges and mutual learning thus made possible have contributed to the progress of human civilization.[5]

A month later he traveled to Indonesia to launch the Maritime Silk Road, a trade and policy framework that would seek to build on existing investment connections between China and Southeast Asia. Reminding the Indonesian parliament about recent bilateral agreements over bridges and dams, Xi Jinping spoke of the much deeper maritime ties between the two countries:

> China and Indonesia face each other across the sea. The friendly ties between us have a long history. Together, our peoples have composed one piece after another of beautiful music about their exchanges and interactions over the centuries. Just as the Indonesian folk song Bengawan Solo, a household musical piece in China, goes "Your water springs forth from Solo, caged by a thousand mountains. Water flows to reach far distances, eventually to the sea." Like the beautiful river Solo, China-Indonesia relations have traversed an extraordinary journey, past mountains and eventually to the sea. As early as the Han Dynasty in China about 2,000 years ago, the people of the two countries opened the door to each other despite the sea between them. In the early 15th Century, Zheng He, the famous Chinese navigator

of the Ming Dynasty, made seven voyages to the Western Seas. He stopped over the Indonesian archipelago in each of his voyages and toured Java, Sumatra and Kalimantan. His visits left nice stories of friendly exchanges between the Chinese and Indonesian peoples, many of which are still widely told today. Over the centuries, the vast oceans have served as the bond of friendship connecting the two peoples, not a barrier between them. Vessels full of goods and passengers travelled across the sea, exchanging products and fostering friendship. A Dream of Red Mansions, a Chinese classic novel, gives vivid accounts of rare treasures from Java. The National Museum of Indonesia, on the other hand, displays a large number of ancient Chinese porcelains.[6]

As during his visit to Kazakhstan, he proposed trade partnerships and exchanges between "youth, think tanks, parliaments, NGOs and civil organizations of the two sides."[7] The concept of reestablishing the overland and maritime Silk Roads for the twenty-first century rapidly captured attention around the world. In the media coverage of the two speeches, debate and speculation centered on the ambition, feasibility, and challenges of undertaking such an initiative. The prospect of oil and gas pipelines, transcontinental rail lines, and deepwater ports led to commentaries on Belt and Road as a geography of connectivity, hubs, and routes—an imagined future conjured by the romance of market towns and the tracks made across mountains and deserts centuries ago. For Xi Jinping, however, the historic Silk Roads evoked more than an infrastructure of long-distance trade; they stood as a metaphor for certain qualities and values, ones that also needed to be revived, as he told his audience in Kazakhstan: "Throughout the millennia, the people of various countries along the ancient Silk Road have jointly written a chapter of friendship that has been passed on to this very day. The over 2,000-year history of exchanges demonstrates that on the basis of solidarity, mutual trust, equality, inclusiveness, mutual learning and win-win cooperation, countries of different races, beliefs and cultural backgrounds are fully capable of sharing peace and development. This is the valuable inspiration we have drawn from the ancient Silk Road."[8]

Since these initial speeches, Xi Jinping has visited dozens of countries, some of them multiple times, to garner support for Belt and Road. The language presented in Astana and Jakarta has been polished and refined, and the idea of reviving a Silk Road spirit has continued to build on ideas and concepts that were central to the party's attempts to steer both domestic and international policies in the previous twenty years or so. Foremost among these have been the concepts of harmony and win-win cooperation. As numerous observers of domestic politics in China have commented, maintaining societal order and suppressing any ethno-cultural tensions in the country have long been overriding concerns.[9] This issue has been particularly charged in the country's northwestern provinces, home to significant populations of Uyghur, Hui, and other minorities. Following these initial launch speeches, the narrative of a Silk Road "spirit" filtered down through the various information management channels of China's international diplomacy. Over the following two years, ambassadors across the region would introduce the goals and principles of One Belt One Road—as it was known in English then—to a wide variety of audiences. Accordingly, in 2015, Qu Zhe, ambassador to Estonia, described the initiative in the following terms: "For thousands of years, the Silk Road Spirit—'peace and cooperation, openness and inclusiveness, mutual learning and mutual benefit'—has been passed from generation to generation, promoted the progress of human civilization, and contributed greatly to the prosperity and development of the countries along the Silk Road. Symbolizing communication and cooperation between the East and the West, the Silk Road Spirit is a historic and cultural heritage shared by all countries around the world and ought to be carried forward. The Belt and Road initiative is the inheritance and development of the ancient land and maritime Silk Road."[10]

This political discourse of an open, outward-looking China, both past and present, took hold with the reform policies of Deng Xiaoping in the 1980s. But as Zhu Zhiqun, Roger Irvine, and others have noted, despite the dramatic changes that have since occurred, inward nationalism and internationalism have remained competing forces in the Chinese political system, a tension that has also led to a strong convergence in how domestic and international affairs are framed and approached. The ongoing emphasis placed on harmony stands as one of the best-known examples of this.[11] From the 1980s, the CCP formulated diplomatic strategies with neighboring countries built around the idea of finding "common interests."[12] Indeed, a terminology of win-win and mutual benefit has underpinned attempts to shift the gravitational forces in Central Asia away from Moscow and toward Beijing. An emphasis on building friendships also reflects Beijing's awareness of the hostilities and anxieties that neighboring countries hold regarding China's past actions and its creeping influence today.[13] The Silk Roads—conceived as Belt and Road—represent a carefully contrived blueprint for addressing such issues. In China's attempts to maintain and foster relations across Central Asia and beyond (hi)stories of cross-cultural fertilization and a dialogue among Buddhism, Islam, Confucianism, and Taoism are fertile grounds for nurturing "friendships" built on "trust" and "mutual understanding." According to Sun Weidong, Chinese ambassador to Pakistan:

> The Chinese Dream is a dream of peace. Throughout history, China has sought peaceful relations with other countries. China will never seize development through colonization and plundering. More than 1300 years ago, the Chinese famous monk Xuanzang came to South Asia through the Silk Road. He stayed in Taxila, which is only 32 km from Islamabad. He learned from and exchanged with others in Taxila, and promoted the friendly exchanges between China and South Asia. 600 years ago, a Chinese navigator named Zheng He led the biggest fleets to South Asia for trade, which stands in sharp contrast to colonization and plundering of the West. Zheng He is a Muslim by the way. China will insist on the principle of peace and friendship, strive for development in a peaceful global environment and promote world peace through self-development.[14]

It is fascinating that Sun imports China's key figure of the Maritime Silk Road, Zheng He, into deliberations about the overland China-Pakistan Economic Corridor to help build bilateral ties around a shared Muslim heritage. It is an example that speaks to the ways in which the five pillars of Belt and Road—most notably people to people ties—have become the architecture through which these framings of culture and history are diplomatically exercised. Chinese politicians have long proclaimed policies of noninterference, where aid and cooperation, they claim, are decoupled from interventions in domestic affairs and proposals for integration or amalgamation.[15] The language of connectivity is thus significant in this regard, where the metaphor of the Silk Road appears to create lines that join proximate and distant places alike in ways that do not challenge the political geographies it connects. The Silk Road thus frames the language of building connections between and among cultures and civilizations, such that difference, it is proclaimed, is both tolerated and respected: "The Belt and Road initiative is an open and inclusive project, not exclusive. It covers, but is not limited to, the area of the ancient Silk Road. . . . It advocates tolerance among civilizations, respects the paths and modes of development chosen by different countries, and supports dialogues among different civilizations so that all countries can coexist in peace for common prosperity," said Qu Zhe, ambassador to Estonia.[16] Moreover, "Chinese leaders have made it clear that in implementing the initiatives, China will uphold the spirit of 'amity, sincerity, mutual benefit and inclusiveness,' which guide China's diplomacy regarding its neighbors, and China will not interfere in other countries' internal affairs, or seek dominance over regional affairs or sphere of influence. The Belt and Road initiatives are for open cooperation, with economic and cultural cooperation being the focus."[17]

Such carefully crafted speeches present a vision of regionalism that differs significantly from the mantra of integrative projects such as the European Union. But while an explicit political coming together is eschewed, the Asian Infrastructure Investment Bank, with its initial $40 billion Silk Road Fund, means that the reality on the ground is somewhat different. Metaphors of connectivity are designed, however, to focus attention on the forms of infrastructure that cross boundaries and link up regions. Camel caravans and the winding pathways of previous centuries are similitudes for oil and gas pipelines, highways, and rail connections stretching across Central Asia into Europe and down to Southeast Asia. Similarly, for the Maritime Silk Road, warehouses and shipyards for fleets of the fifteenth century are the evocative images for building today's network of deepwater ports, container shipping terminals, and massive coastal developments. The significance of this language lies in China's future energy-security needs and ambitions for long-term economic growth. As Devonshire-Ellis and others have highlighted, the construction of an infrastructure matrix across the Eurasian region, driven largely by Beijing and Chinese firms operating internationally, represents a massive antirecessionary stimulus package for the coming decades.[18] In a book examining the Eurasian energy sector published in 2012, Kent Calder suggests that China's standing as the region's largest energy-insecure consumer economy is a catalyst for the political economy of the region and producing a new form of continentalism. Writing more than a year before the launch of Belt and Road, Calder insightfully points to the Xinjiang region as a "continentalist growth pole."[19] This, together with China's need to reduce reliance on potentially volatile corridors such as the Straits of Melaka and the Turkmenistan-Uzbekistan route, are considered in greater detail in the subsequent chapter.

Among the most prominent architects of China's foreign policy in the aftermath of Belt and Road's launch was Yang Jiechi. In his capacity as state councillor, a role that provides oversight to various government departments, Yang delivered a series of high-profile speeches outlining the scope of Belt and Road to international audiences. The following excerpt from a presentation in 2014 to the increasingly influential Boao Forum for Asia vividly captures the subtle ways a historical Silk

Road has come to be invoked, as well as how the mantra of peaceful, cultural connections is used to build trust within Eurasia's fast changing and multipolar political landscape:

> The Silk Road has given the people of Asia confidence to pursue inclusive cooperation. The Silk Road had enabled the East and the West to thoroughly interact with each other in peace and equality in all possible areas. The network for trade running over the Eurasian continent way back in the early days brought benefits to all sides. The ancient Chinese technologies of iron-smelting, farming and irrigation were brought to neighboring countries to the west and south of China and greatly increased those countries' productivity, while the medicine, calendar and sculpturing art from South Asia were applied and valued in China. The Chinese Harp and Polo, created by ancient nomads in Central Asia, featured proudly in the history of Chinese music and sports. The sandalwood from Timor-Leste, elephant from Thailand, wood sculpture from Laos, leather and fur from Russia, colored glaze from Europe and west Asia, and gems from Pakistan, Myanmar and Sri Lanka were all highly popular with ancient Chinese. Some religions or sections of religion, after being introduced into China via the land and maritime silk roads, integrated with homegrown religions in China and coexisted with them in harmony. For the numerous Chinese and foreign envoys, merchants and the wise questing for scriptures and knowledge, it was the Silk Road, and the people who kindly assisted them along the way, that had made their journeys possible.
>
> Now that conditions for transport and infrastructure have much improved and regional cooperation has entered a new stage, countries in Asia may well live up to the Silk Road spirit of peace, friendship, openness, inclusiveness and win-win cooperation, and endeavor to add a new dimension to the Silk Road spirit.
>
> An important part of this new dimension, as I see it, is mutual trust. The Chinese people believe that "one would achieve nothing without credibility." In fact, travelers on the ancient Silk Road could hardly make the journey alone. They had to travel in company and look after each other on the way. They even had to work with each other when they sailed in a same boat on the sea. Despite the leapfrog development in science and technology, cooperation in transport today still faces hidden difficulties, and some Asian countries still suffer from a "deficit of trust." I believe as we work to improve connectivity and promote the Belt and Road initiatives, more needs to be done to increase mutual trust. What forms the basis of mutual trust is for countries to respect history and draw lessons from it, follow a path of peaceful development.[20]

As we will see, there are geographies to this language that are made and remade by different actors describing connectivity from different locations. On occasions, the two Silk Roads are described as a geography of seemingly boundless and inclusive connections. Xi Jinping, for example, has proposed that "the sea is big because it admits all rivers" to signal that the Maritime Silk Road is an open and inclusive initiative.[21] But elsewhere a terminology of *gateways* or *starting points* and *ends* has been used to affirm the strategic interests of those invested in defining the geographical parameters of the initiative both within and outside China. Most explicitly, speeches from senior Chinese diplomats and politicians have sought to firmly anchor the origins of the Silk Roads, in both its dates and locations, within China itself. At the opening keynote of the Belt and Road summit in May 2016, Zhang Dejiang stated: "Over 2,000 years ago, there were already interactions between the two ends of the Eurasian continent, and that was when the ancient Silk Road began. The road had taken shape by the time of the Qin and Han dynasties of China, and was most prosperous during the Sui and Tang dynasties. During the Tang and Song dynasties, our ancestors developed advanced navigation techniques, and successfully opened up the Silk Road on the Sea that extended all the way to the West."[22] Earlier, at the Boao Forum for Asia in April 2014, Yang Jiechi commented: "Europe was the end of the ancient Silk Road, yet silk originated from China, where the Silk Road started, and many important areas along the Silk Road were in Asia. The Asian people, the Chinese people included, opened the Silk Road, withstanding great hardships, and preserved it throughout the years. We therefore feel a natural affinity for the Silk Road, which had once witnessed the common history and glory of the Asian civilizations. For the Asian people, the Silk Road provides a source of historical and cultural pride, and stands as a flag of unity and cooperation among Asian countries."[23]

Locating the eastern end of the overland route in China, and specifically at Xi'an, effectively excludes countries to the east, most notably Japan, from the strategy for Eurasian physical and economic integration. Most contentious, though, encapsulating the maritime connections between China and Southeast Asia in the story of Zheng He feeds into the disputes and tensions over the South China Sea. To set the scene for a discussion of the different layers of the Zheng He issue presented in the following chapters, it is first helpful to briefly trace the origins of his appropriation in recent Chinese diplomatic and political discourse, and how Belt and Road elevates the narrative in new ways. Fleetingly cited by Deng Xiaoping back in 1984 as an early example of an open-door policy, Zheng He came to prominence as a form of cultural propaganda in the late 1990s as part of an increasingly internationalist China. During a speech at Harvard University in late 1997,

Jiang Zemin cited Zheng He, among other early travelers, to suggest that China has long been open and has spread the influence of its culture far and wide.[24] Under Hu Jintao (2003–2013), Zheng He would be more explicitly presented as a figure of peace and friendship. In speeches given in Indonesia, at the Arab League of Nations in Egypt, and at the University of Cambridge, the Chinese premier Wen Jiabao juxtaposed the country's present-day ascendancy to the image of a benign maritime heritage: "In the 15th century, the famous Chinese navigator Zheng He led seven maritime expeditions to the Western Seas and reached over 30 countries. He took with him Chinese tea, silk and porcelain and helped local people fight pirates as he sailed along. He was truly a messenger of love and friendship. The argument that a big power is bound to seek hegemony does not apply to China. Seeking hegemony goes against China's cultural tradition as well as the will of the Chinese people. China's development harms no one and threatens no one. We shall be a peace-loving country, a country that is eager to learn from and cooperate with others. We are committed to building a harmonious world."[25]

Jeff Adams offers a detailed account of this diplomatic use of Zheng He's voyages in the context of China's growing naval powers and competing national interests concerning the South China Sea, the Straits of Melaka, and the Java Sea. With China in a militarily weak position relative to the United States, Zheng He provides a valuable soft-power resource with regional neighbors. In the late 1980s, the first ship to be designed and built by the People's Liberation Army Navy began training voyages, sailing around the world. With a crew of around two hundred, the 433-foot *Zhenghe* continues to propagate the idea of China as a benign maritime power, as reports of an upcoming trip to Southeast Asia and the southern waters beyond in late 2016 indicates: "As a 'floating territory of the country,' Zhenghe ship will send Chinese people's friendly greetings to the people of the visited countries to deepen the friendship and display the new image of the constantly-developing Chinese Navy. . . . Since its launch nearly 30 years ago, Zhenghe ship has been to the waters and ports of 28 countries in its 13 voyages, carrying the envoys of friendship from Chinese naval colleges and universities."[26]

Shifting the Geographies of Internationalism

Such maritime diplomacy sits in stark contrast to the tensions that have steadily grown since 2010 regarding the South China Sea and its territorialization. The construction of airfields and bases on the Spratly and Paracel Islands, together with territorial enforcements by the Chinese navy, greatly increased suspicions toward Beijing and created open hostility over maritime boundaries and fishing and oil rights in the region. In 2012, the Scarborough Shoal emerged as a flash point between China and the Philippines, with disputed claims of ownership leading to protests in both countries and a convoluted game of chess as governments around the world reacted with statements of support, or with more diplomatically careful calls for multilateral resolutions.[27] As tensions over the South China Sea issue continued to rise, the ruling by the International Court of Justice in The Hague in July 2016 rejecting China's claims of historical rights to the area within the "nine-dash line," in favor of the Philippines, represented a landmark moment. But with Beijing denouncing the decision, diplomatic pressure mounted for it to scale back its island construction program. Vietnam maintained its defensive stance, fortifying its island bases with rocket launchers in the disputed Spratly Islands, and the US and Chinese militaries continued to flex their muscles through exercises in disputed territories, provoking condemnations from the other side.[28] The arrival of Donald Trump in office in early 2017 further escalated tensions, as Washington think tanks such as the Center for Strategic and Budgetary Assessments called for a hardening of relations with China.[29] It was against this backdrop that the diplomatic dances of the new Maritime Silk Road played out. As the territorialization of the South China Sea threatened to overshadow Belt and Road, relations between China and ASEAN (Association of Southeast Asian Nations) and various countries in Southeast Asia remained fragile. In this context, the "legend" of Zheng He regularly featured in China's attempts to smooth out relations with its regional neighbors. Xi Jinping's visit to the National University of Singapore in 2015 offers a case in point:

> China and Singapore are friendly neighbors across the sea with a long history of amicable exchanges. In the early 15th century, China's great navigator, Zheng He, called Singapore several times on his ocean voyages. A full size replica of the treasure boat of Zheng He is on display in the maritime museum of Singapore to honor this historic event. In the late Ming and early Qing dynasties, many people from China's Guangdong and Fujian provinces migrated to Southeast Asia, bringing with them Chinese culture and skills, and sowing the seeds of China-Singapore friendship. . . .
>
> Since the ancient times, we Chinese have valued harmony in diversity and good-neighborliness, which have much in common with the values of the people of

Southeast Asia. The Chinese culture cherishes such values as benevolence, virtue, modesty, self-reflection, learning, and pursuit of excellence. In many Chinese literary classics such as the Romance of the Three Kingdoms and All Men Are Brothers, the protagonists are both loyal and righteous, and these are the qualities that are admired by the Southeast Asian people as well. Similarly, the cuisine, music, architecture, painting and religions of Southeast Asia have also influenced the Chinese culture. . . .

In modern times, China and Southeast Asian countries encouraged and supported each other in the cause of independence and liberation, and we have inspired and worked with each other in economic and social development. . . . China's neighborhood occupies a top priority on its diplomatic agenda, and China has the unshirkable responsibility to ensure peace, stability and development in its neighborhood. China is dedicated to promoting a more just and equitable global governance system, enhancing democracy in international relations as well as the building of a new type of international relations based on win-win cooperation and a community of shared future for mankind. Efforts to reach this goal should naturally start in its neighborhood.[30]

With Zheng He portrayed as an envoy of peace and friendship through explicit and implicit inferences, his story is misleadingly offered as a metonym for a long history of trade, cultural exchange, and coprosperity, and thus stands in for China's maritime intentions. Chapter 4 explores the distortion of the Zheng He narrative in greater detail. But it is also evident that at a more subtle level Zheng He forms part of a Maritime Silk Road discourse that seeks to shift the focus of the South China Sea and the Straits of Melaka away from disputed ownership rights and toward a language of routes, connections, and free passage. In this regard, claims of benign connectivity act as a diplomatic countervail to the seemingly intractable and cyclical hostilities that surround remote island disputes, patrolled borders, and the territorialization of "open waters":[31]

The Chinese philosophy values peace as being the most precious. The Chinese history shows a record of China in friendly relations with neighbors. And China's diplomacy honors the tradition of matching words with deeds. Former Malaysian Prime Minister Datuk Seri Mahathir observed that when Zheng He, the Chinese navigator of the Ming Dynasty, led the most powerful fleet to Southeast Asian countries more than 600 years ago, he brought along good things and genuine friendship, totally different from Western colonizers that came afterwards. Today, the people of Southeast Asia still cherish a fond memory of Zheng He. In fact, the "Zheng He Association" is going to be set up in Malaysia to promote trade and investment between Malaysia and China. The Belt and Road initiatives China put forward fully reflect the commitment to mutual trust and mutual benefit.[32]

Looking beyond the contexts of Indonesia and Malaysia, Zheng He forms part of a narrative of peace that Beijing has been projecting globally. China has faced sustained criticism concerning its activities and investments in Africa and Southeast Asia. Evidence of rising nationalism, together with increased military spending and bullying of neighbors, has also created significant anxiety in countries across Asia-Pacific and beyond. The diplomatic narratives that underwrite Belt and Road thus represent an attempt to project the image of a rising peaceful power, one that is increasingly declaring intent to offer greater leadership in international affairs and multilateral government structures.[33] To that end, Zheng He has been regularly cited alongside examples of UN peacekeeping missions and military humanitarian-aid interventions as evidence of China acting as a good citizen in the international order.[34] The precise directions of China's foreign policy under Xi remains unclear, in part because of the competing worldviews held within the CCP.[35] Nonetheless, Belt and Road clearly signals Beijing's commitment to a foreign policy that couples a more confident and assertive nationalism with internationalist modes of development.[36] To interpret this approach, Irvine cites the Institute of Modern International Relations at Tsinghua University as an increasingly important source of intellectual guidance. He suggests that a new "Tsinghua approach" to Chinese international relations theory, which partly draws on ancient Chinese thinking, has begun to shape policy making.[37]

In the example of Greece noted earlier we saw a co-opting of the Silk Roads to construct ideas about the reclaiming of dignities after episodes of humiliation. Although this is typically articulated in the context of bilateral discussions, the following excerpts reveal how such a strategy has been rolled out across the region:

In recent history, both China and Pakistan suffered from imperialist and colonialist aggression and oppression. Similar historical sufferings and the common struggle have brought our hearts and minds together. Since the establishment of diplomatic ties, our two countries have forged an all-weather friendship and pursued all-round cooperation, and we have supported each other on issues crucial to our respective core interests. Having gone through weal and woe together, we couldn't feel more gratified to have each other as great neighbor and friend. . . . We should advance our shared interests and achieve common development. We should use China-Pakistan Economic Corridor to drive our practical cooperation with focus

on Gwadar Port, energy, infrastructure development and industrial cooperation so that the fruits of its development will reach both all the people in Pakistan and the people of other countries in our region.
XI JINPING, APRIL 2015

Our two peoples sympathized with and supported each other in their respective struggle for national independence and liberation in the last century. Indonesia was among the first countries to establish diplomatic ties with the People's Republic of China after its founding in 1949. In 1955, China and Indonesia, together with other Asian and African countries, jointly initiated the Bandung spirit at the Bandung Conference. With the principles of peaceful coexistence and seeking common ground while shelving differences at its core, the Bandung spirit remains an important norm governing state-to-state relations, and has made an indelible contribution to the building of new international relations.
XI JINPING, OCTOBER 2013

Over 40 years ago, some 50,000 Chinese men and women journeyed to Africa, where they shed sweat and blood to build Tazara, a 1,860-kilometer railway linking Tanzania and Zambia on the African continent. That was the time when things were pretty tough for China itself. Yet, by tightening our own belt, we managed to lend a helping hand to a large number of developing countries in Asia, Africa and Latin America to assist them in their struggle for independence and liberation. The reform and opening up program that started over 30 years ago has enabled us to marry China's advantage in market and labour resources with the capital and technology of developed countries in the pursuit for win-win cooperation.
FOREIGN MINISTER WANG YI, MARCH 2015[38]

Not surprisingly, this language of mutual struggle and prosperous futures was rapidly adopted by a number of Belt and Road countries, with Iran, Serbia, and Malaysia among those having officially acknowledged Beijing's empathies as a sign of goodwill.[39] The significance of a Silk Road geography, then, is that it elevates such affinities to a regional scale and harnesses a discourse of international cooperation that seemingly sidesteps religious and political tensions in an attempt to build networks that incorporate the Arab world and Iran, countries across the African continent, as well as smaller economies as far apart as the Mediterranean and Southeast Asia.

The voices we have heard so far are those of Chinese diplomacy and foreign policy. But as the language of a Silk Road revival has been exported across Eurasia, East Africa, and beyond, governments jockeying for a favorable position in the emergent network of bilateral investments have signaled their own presence in the diplomatic genealogies of the Silk Roads. As noted in chapter 1, Sri Lanka has rapidly emerged as a key strategic point on the Maritime Silk Road. In the aftermath of a prolonged civil war, the Rajapaksa government saw Belt and Road as a lucrative opportunity for developing the country's infrastructure and manufacturing and tourism sectors. To help realize ambitions of strategically positioning the country in the Indian Ocean shipping industry, the government seized on histories of Ming dynasty exploration. As we will see in chapter 6, this involves glossing over violent battles and Zheng He's kidnapping of Sri Lanka's king in favor of a more diplomatically expedient language of encounter, exchange, and harmonious trade. Belt and Road has presented Sri Lanka with an unprecedented opportunity for economic development and foreign investment. As this diplomatic entanglement of the cultural past into foreign policy has been polished since the launch of BRI, the Sri Lankan government has also increasingly drawn on the rhetoric of peace, friendship, and security to project grand ambitions for the coming decades, as the following speech from early 2016 by Prasad Kariyawasam, ambassador to the United States, illustrates:

> Sri Lanka has been known to the travelers of the ancient world as a hub in the Indian Ocean, who identified the country with many names like Lanka, Serendib and Ceylon. . . . The people of Sri Lanka, as islanders, since ancient times, have been influenced by several waves of external interactions that led to the exchange, not only of goods but also ideas and knowledge with travelers and traders passing through or visitors from lands close and far. . . . As a result, Sri Lanka is a multi-ethnic and multi-cultural nation. Sri Lanka benefited from the message of the Buddha who lived and taught in Northern India. Arab traders brought with them the teachings of Prophet Mohammed. It is believed that Guru Nanak, the founder of the Sikh religion, visited Sri Lanka in 1511. . . .
>
> It is clear that the world needs peaceful oceans to sustain its benefits in the ever growing blue economy. . . . Sri Lanka has the benefit of the vast ocean around us over which we enjoy exclusive economic rights. The country is also situated on the world's busiest shipping lane. This is both an opportunity and a challenge for our nation, which is situated right in the middle of the Orient between East and West. . . .
>
> Sri Lanka takes the security of sea lanes and maritime security in the oceans around us, seriously. We are eager to work with the maritime powers of the Indian Ocean and beyond, to make our oceans secure for unimpeded commerce and peaceful navigation. We are determined, as it is in our interest, to work with the maritime powers of the region to ensure that the Indian Ocean is conflict free. We

will therefore work with the maritime powers of the region with commitment to prevent conflict, combat terrorism and piracy and assist to harmonize geo-strategic complexities. . . .

There are several hubs in the world that we recognize as important: New York, Dubai, Singapore, Hong Kong and others, all with bustling ports nearby. Centered on the City of Colombo and its deep water Port, Sri Lanka too can assume such a status. Sri Lanka intends to broaden the existing Indo-Sri Lanka Free Trade Agreement . . . Sri Lanka enjoys a Free Trade Agreement with Pakistan, and is now working towards such an Agreement with China.[40]

In a similar vein, for those in Malaysia looking to capitalize on closer ties to Beijing, Zheng He's voyages have been co-opted to signal the country's legitimacy in a Sinocentric history of globalization:

History tells us that back in the 16th century, when China found out that Portugal had invaded Malacca, the Emperor of the Ming Dynasty was deeply offended. According to the well-known Portuguese adventurer Tomé Pires, Portugal's application for the establishment of an embassy in China was then rejected in 1521. Two years later, in retaliation for the invasion, the Chinese authorities executed 23 Portuguese and incarcerated many more. The point of this story is that Malaysia and China have a common history that goes back many centuries. The 15th and 16th centuries essentially tell us about the era of "globalisation" as spearheaded by China then—when the word or the concept was not even heard of. The Ancient Silk Road as a prototype version of globalisation goes back even further to the 2nd century BC when during the Han Dynasty of China, an elaborate network of trade routes connected major cities in the West and the Middle East to China.

Today, we have the Maritime Silk Road which I would regard as an alternative road to globalisation in the 21st century in the sense that it is not a "West-centric" paradigm. To my mind, this new Silk Road offers a more "Asia-centric" approach in trade and commerce, and cultural ties and opens up vast opportunities for multilateral dealings not just between China and ASEAN but the rest of Asia as well.[41]

Since the launch of Belt and Road, governments across Eurasia and Africa have turned to a language of Silk Road connectivity. Although references to cooperation, trust, and friendship have always been the mantra of diplomats and the propaganda of politicians engaged in foreign policy, these have come to be accompanied by the language of win-win and coprosperity by those jockeying for position in the new world of Belt and Road. The lesson of history, they have repeatedly told their audiences, is that the dividend of open borders and cross-border trade is affluence and harmony. Fascinatingly, the question of where a country's ambitions lie in the unfolding geographies of Belt and Road—overland or maritime—leads to the deployment of different metaphors of a Silk Road past. For Sri Lanka and Malaysia, the ancient Silk Road is about hubs and nodes. These have material connections to the present day with both countries hoping to secure investments in deepwater ports and container shipping terminals in an increasingly dense network of oceanic shipping. In contrast, countries on the overland route such as Pakistan and Georgia frame their experience as a story of routes and links, in alignment with their twenty-first-century ambitions to build rail and roads, as well as energy pipelines.[42]

The story of Belt and Road will continue to primarily be about the bilateral relations China can build with its partner countries. The tropes of friendship, trust, win-win cooperation, and mutual benefit speak to these futures. Before moving on to consider how such ideas are materialized through particular corridors of cooperation and development, I offer one final excerpt, one that tellingly reveals the larger ambitions underpinning the Silk Roads revival discourse. During the 2014 Boao Forum for Asia, Yang Jiechi invoked the past again, this time elevating it to a template for future coprosperity for a "rising Asia": "Ladies and gentlemen, the history of Asia has evolved in close relation with the rise and fall of the Silk Road. The collective renewal of Asia in the future will come alongside the revival of the Silk Road. The Belt and Road initiatives will have an overall bearing on the big Asian family. We hope that countries concerned may discuss and work with China as these two initiatives are being implemented, which will bring benefit to all of us. Let us work together with real earnest."[43]

FOUR

Corridor Diplomacy

Belt and Road is in large measure an exercise in tidying up. Major infrastructure projects linking cities, oil refineries, economic trade zones, and container shipping ports across Asia have been in development for decades. The discursive leap was to place China at the center of this regional integration and give it a unifying, overarching vision of win-win cooperation. As we have already seen, the language of Belt and Road is one of ambitious acceleration and scaling up. This situation is largely mirrored in the leveraging of history and heritage. Belt and Road has not engineered heritage diplomacy into existence. Indeed, as with other aspects of the initiative, some preexisting patterns have been reframed and given a whole new impetus and scale of significance. As a result, new players have emerged and strategies have been amplified and fine-tuned. The examples offered in this chapter highlight some of the foundations on which Belt and Road heritage diplomacy has been built and the continuities that exist in the years before and after its 2013 launch. At the same time, it also suggests that Belt and Road has, in some cases, qualitatively changed the game, whereby the mirage of the ancient Silk Roads alters how space is thought about, how international relations are conducted, and how trade networks are configured. Building on the themes of the previous chapter, the focus is again on the links among economic development, heritage preservation, tourism, and discourses of culture, observing how such connections come to be made via the planning of World Heritage Sites, museums, and cities. Particular attention is paid to Xinjiang and the likely implications of cultural-sector development for local communities. Here world heritage is analyzed alongside Uyghur music traditions to consider some likely future scenarios of heritage politics in an area critical to the success of Belt and Road. In this respect, events of today are located within the broad trends of the twentieth century of state governance in the region. With this followed by an analysis of maritime issues, the two Silk Roads are discussed separately to substantiate the argument that Belt and Road has become a platform on which particular transnational corridors of heritage production are now forming.

As noted in chapter 1, to introduce a degree of specificity to the broad geographies of the overland and maritime components of Belt and Road, China's Ministry of Foreign Affairs and Ministry of Commerce devised the Vision and Actions Plan in 2015, featuring five economic corridors and an overarching Eurasian land bridge. Delineating the region in this way clearly speaks to a long-term investment strategy in the hard infrastructure of cross-border trade, economic development, and energy provision. In that regard, the concept of the corridor is also about linking up "key economic industrial parks" and cities as far apart as southern Europe, eastern China, and Southeast Asia, with multiple urban nodes of connectivity in between.[1] The introduction of five "cooperation priorities" in the Vision and Actions Plan—policy, facilities, trade, finance, and people-to-people—mean that initiatives launched along these corridors are also reaching across a broad range of social and cultural fields, including medicine, education, security, tourism, cultural and environmental preservation, traffic management, optical cable networks, civil aviation, banking, and film festivals, to name a few. In the wake of the various announcements concerning Belt and Road since 2013, the list of initiatives in the cultural sector also began to grow rapidly. In 2014 the Silk Road International Cultural Forum scheme was launched, with an initial event in Astana, Kazakhstan, followed by a second forum in Moscow a year later. The Silk Road International Arts Festival expanded its ambitions rapidly, with its third annual event in 2016 in Xi'an featuring calligraphy, painting, and photography exhibitions, as well as two weeks of performances by artists from more than eighty countries. The year 2014 also saw the launch of two rival Silk Road film festivals, the first in Dublin, and then in Beijing, which commenced its annual program in October. A year later the China Silk Road Foundation initiated a collaboration with the United Kingdom–based 1001 Inventions for exchanges and events based on the history of scientific discoveries and inventions of Silk Road countries. In July 2015 the inaugural Silk Road Dunhuang

International Music Forum took place in Gansu province. The strategic benefits of holding the event in Dunhuang were outlined by the program's executive secretary-general:

> Guansu [*sic*] is where the land Silk Road begins. The ancient Silk Road was a vital communication line to the west, Dunhuang is located at the major crossroad of the Silk Road. It is the trade centre and transit hub between China and the West. Gansu occupies an important status in China's current "One Belt and One Road" Strategy. . . . In order to implement the strategy, a major cornerstone needs to be established first, which is how China's culture and values can win recognition from the countries along the Silk Road, how to win bilateral and multilateral trust and cooperation based on such recognition. Dunhuang has a rich culture, especially the culture of music, music has no boundary of space and time, it is one of the important elements to establish such a cornerstone. In August 2014, UNESCO and famous Chinese musician Tan Dun joined hands and held a grand concert—the Map of Symphony: New Silk Road—Chang'an. Tan Dun said during the concert, the musical elements of the Silk Road and thousands of years of Chinese history were fully blended, starting in Dunhuang in Gansu, integration and exchange of music will undoubtedly bring our grand goal into the next level.[2]

Less than a year later, the Silk Road International League of Theaters launched in Beijing as a collaboration among sixty-six theaters located across twenty-one countries. More than three hundred delegates from Europe and Asia also attended the inaugural Development Cooperation Forum on International Industrial Parks and Silk Road International Culture Week in Xi'an.

Similar cultural cooperation programs launched in various countries for the Maritime Silk Road. In Hong Kong, the *One Belt One Road Visual Arts Exhibition* featured artists from countries around the region and included fashion shows with clothes "inspired by the ancient Silk Road's role in cultural exchange."[3] To help raise its visibility as a key hub of maritime trade during the Yuan dynasty, the city of Quanzhou hosted its inaugural Asian Arts Festival in late 2015, involving performers and delegates from twenty countries. In summarizing the strategic value of the event, a representative of the organizing committee indicated that "the arts festival this year will echo China's Belt and Road Initiative. Models of cultural exchanges along the ancient route of the Maritime Silk Road will be explored during the festival, also to stimulate Fujian's cultural trade in the future."[4]

It would also not take long for the tentacles of Silk Road heritage diplomacy to reach the United Nations, with New York headquarters hosting theatrical performances and art exhibitions on the Maritime Silk Road less than three years after Xi Jinping's first speeches in Kazakhstan and Indonesia. A collection of oil paintings by the artist Feng Shaoxie re-created historical scenes from Turkey, Malaysia, Macau, Hong Kong, and southern China. In many ways, then, Belt and Road heralded a new era of cultural cooperation and cross-border dialogue. Equally significant, however, it also provided a new political economy for a number of preexisting programs and trends, elevating their significance and momentum. To illustrate this, the body of this chapter focuses on the themes of world heritage and Zheng He in relation to broader trade and infrastructure projects, as well as the political and economic entanglements driving this wave of international cooperation in culture and history.

Corridors of Silk

Clearly, Belt and Road dramatically changed the political impetus for cultural-sector cooperation in Asia. In addition to the examples already noted, the evidence for this can be found in the distinct upturn in activity in intergovernmental collaborations for the nomination of historic Silk Road sites to UNESCO's World Heritage List. Both UNESCO and the UN World Tourism Organization have long recognized the need to give greater international attention to the historical and cultural connections and flows of the overland Silk Road, the legacy of which is an array of movable and immovable material culture, including monasteries, post houses, forts, cave temples, Buddhist pagodas, tombs, textiles, documents, leather goods, and ceramics.[5]

Nominating archaeological sites or individual buildings to the World Heritage List for their significance as part of a Silk Road history presents several challenges. Given that many are located in remote regions or in countries that lack experience in fulfilling UNESCO's expectations for both nomination and post-listing management, there are significant bureaucratic and technical hurdles that need to be overcome.[6] Perhaps more significant, a shift in mind-set in how sites are ascribed value is required. Throughout Asia, states have historically identified the finest artistic and architectural achievements found within their borders as the patrimony of the nation. The invention of modern cultural heritage industries in postcolonial countries has been particularly charged in this regard where the forging of identities and a sense of cultural and political sovereignty have defined the narratives of the past. Identifying the cultural heritage of the Silk Roads, both maritime and overland, thus requires states to look beyond their own borders and in part to relinquish the concept of *patrimoine national* in favor of a language of shared heritage, one that is based on itinerant histories. Although experts in fields like conservation and archaeology might freely advocate such approaches, political will and resources are required to enact this into policy. Of course, no single country can achieve this on its own; partners with mutual interests and the suitable historical connections need to be found, and the appropriate institutional relationships need to be built. In the past decade or so, the World Heritage List has begun to reflect this approach, with a steady increase in the number of multisited "serial nominations" that bring together two or more countries in a collaboration. The example of Qhapaq Ñan, Andean Road System, illustrates how this can hold diplomatic benefits for those involved. The listing of a network of Inca roads and structures spanning thirty thousand kilometers in 2014 occurred on the back of a ten-year multilateral collaboration involving six South American countries. Beneficial for cross-border tourism, the project generated significant goodwill between the countries involved and toward the government of Spain, which sponsored the decade-long cooperation (fig. 4.1). Such projects reflect a wider trend in the World Heritage system, whereby

4.1 World Heritage listing celebrations for Qhapaq Ñan, Andean Road System, Doha, 2014. Photo by author.

representatives from states parties contributing to discussions at the World Heritage Committee's annual meetings liberally use terms such as *flows, crossroads, cultural ties, bridges,* and *pilgrimage* to justify inscriptions on the World Heritage List.[7]

After several years of discussion, plans for listing sites of the overland Silk Road formally commenced in 2003. It would be some years later still before more-regular meetings created the momentum necessary for preparing nomination dossiers. To assist the process, the Intergovernmental Coordinating Committee of the Serial World Heritage Nomination of the Silk Roads was established in 2009, an initiative made possible through the financial support of the Japanese, Norwegian, and South Korean governments. Meetings in China, Turkmenistan, and Kyrgyzstan between 2009 and 2012 led to the submission of two world heritage nominations to the world heritage office in Paris. Both were considered at the 2014 meeting of the World Heritage Committee in Doha, Qatar. The joint submission by Tajikistan and Uzbekistan for the Penjikent-Samarkand-Poykent Corridor was referred, with requests for more detailed supporting documentation. The second nomination was more successful, with China, Kazakhstan, and Kyrgyzstan jointly celebrating the addition of the Silk Roads: The Routes Network of Chang'an-Tianshan Corridor to the prestigious World Heritage List.[8] The dossier included thirty-three separate sites, twenty-two in China, eight in Kazakhstan, and three located in Kyrgyzstan (fig. 4.2). On the back of this success, UNESCO's project officer for the Silk Roads, Roland Lin Chih-Hung, suggested the inscription held significant potential for developing national capacity across the region and would stand as a model for the preparation of future serial and transnational nominations around the world.[9] Both nominations were guided by a comprehensive report authored by Tim Williams on behalf of the International Council on Monuments and Sites, known as ICOMOS. Published in its final form in 2014, *The Silk Roads: An ICOMOS Thematic Study* detailed 550 historic settlements, archaeological sites, and architectural structures across the region associated with the Silk Road, grouping them into a series of transnational corridors for the purposes of World Heritage Site listing (fig. 4.3).[10] Of these, around 220 appear on the tentative lists produced by states across the region, the vast majority of which, Williams suggests, could have their significance reframed to tie them into future Silk Road nominations. Thirty-five sites are already inscribed as World Heritage Sites but could be renominated as Silk Road properties, joining the thirty-three that were included in the

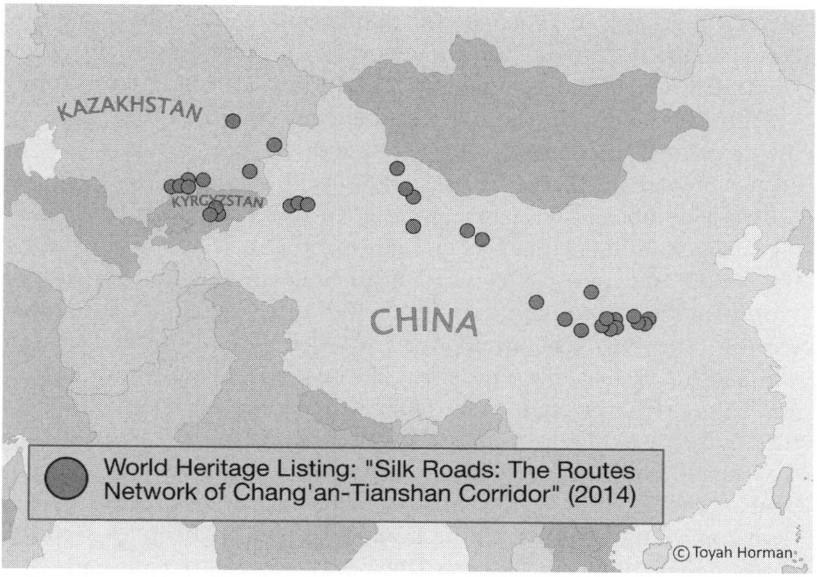

4.2 Map of thirty-three properties in World Heritage listing for "The Silk Roads: The Routes Network of Chang'an-Tianshan Corridor." Adapted from Tim Williams, 2014. Courtesy of Toyah Horman.

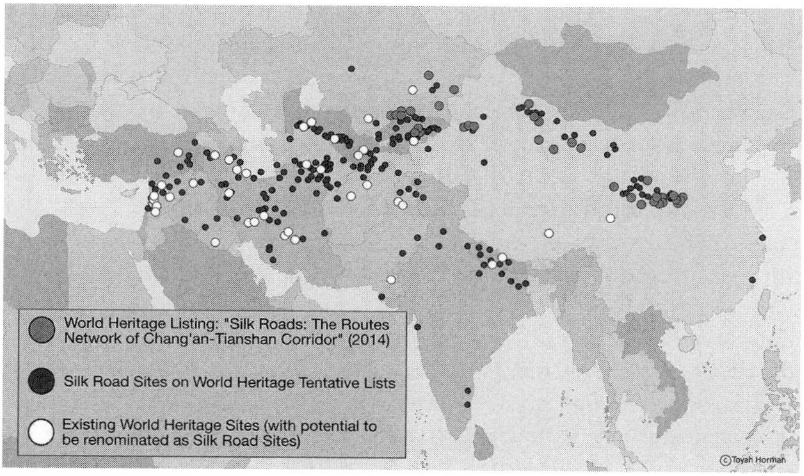

4.3 Locations of sites on World Heritage and Tentative World Heritage Lists. Adapted from Tim Williams, 2014. Courtesy of Toyah Horman.

successful Chang'an-Tianshan Corridor nomination. The remaining 260 or so are documented in the report as sites that do not appear on any UNESCO lists but hold potential for future Silk Road submissions.

From 2013 onward, the fourteen member countries that made up the coordinating committee began to address such possibilities with greater regularity.[11] Between 2013 and 2016, Kazakhstan, for example, hosted a number of conferences designed to advance Silk Road heritage cooperation. Such events brought together delegates from around the region and from UNESCO to advance documentation for further World Heritage nominations.[12] Funds to directly support these initiatives and the long-term development of collaborations through Central and South Asia have been provided by the Japanese and South Korean governments. Over eight years Japan donated nearly $1.7 million to the Silk Roads program, with Korea initiating its support in 2013 with a $200,000 contribution.[13] In late 2015, Nursultan Nazarbayev, president of the Republic of Kazakhstan, signaled his government's long-term commitment to promoting heritage as a mechanism for security, peace, and stability. Speaking at UNESCO's headquarters in Paris, Nazarbayev indicated that his country would take a leading role in this area via the new Academy of Peace, an initiative that was subsequently renamed the Center for the Rapprochement of Cultures, in alignment with UNESCO's 2013–2022 program of the same title. Initial planning meetings for the center indicated how its activities would contribute to "the revival of the spiritual and cultural belt of the Silk Way."[14]

Significantly, this wave of activity in creating international collaborations around culture and heritage, and the desire by Kazakhstan to play a leading role therein, directly maps onto the geographies of Belt and Road. As figure 4.4 illustrates, the various Silk Road sites in Central and South Asia under consideration for listing as World Heritage corridors overlay the routes Beijing has identified for trade and infrastructure investment, most notably the corridors of Bangladesh-China-India-Myanmar and of China–Central Asia–West Asia. Since the launch of Belt and Road, agreements have been signed for multibillion-dollar Chinese investments in the development of deepwater ports in Chittagong, Bangladesh, and Kyaukphyu, Myanmar.[15] Both countries have become key strategic partners in Beijing's plans to reduce the costs and risks of shipping oil and gas from the Middle East. Their locations in the Bay of Bengal mean tankers and container ships can avoid the long and potentially perilous route via the Straits of Melaka and South China Sea. Since coming to office, Xi Jinping has stepped up China's investment in Bangladesh. Major funding and assistance for

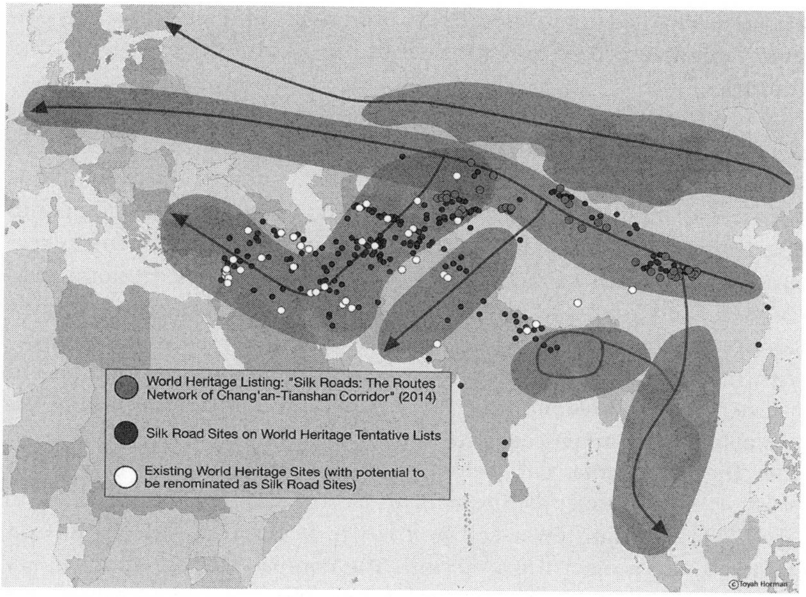

4.4 Overlay of sites for Silk Road World Heritage listing and Belt and Road economic corridors. Courtesy of Toyah Horman.

building bridges, airports, rail lines and pipelines, power plants, and port facilities all speak to a strategy of securing an infrastructure corridor into southern China and new shipping lanes across the Indian Ocean.[16] In Pakistan, China has committed to a long-term investment in infrastructure development estimated in the region of $50 billion. Loans are expected to exceed $10 billion over the coming years, of which $3.7 billion has been committed to the construction of a rail link between the country's major cities.[17] If the construction of a deep-water port and road, rail, and pipeline connections in the southern port city of Gwadar are completed, the city will provide China with an additional strategic access point, this time to the Arabian Sea.[18] Belt and Road has also led to a significant increase in Chinese assistance to Kazakhstan, with loans funding an ambitious infrastructure program involving thousands of miles of road and rail. With the government in Astana strategically positioning the country as a trade gateway between eastern and western Asia, in early 2015 China announced agreements for thirty-three infrastructure collaborations, worth more than $23 billion.[19] These were followed by further deals signed later in the same year estimated on the order of $10 billion.[20] Similarly, in Kyrgyzstan,

China has loaned approximately $1 billion to build a north-south corridor designed to link up with Belt and Road projects in neighboring countries.[21]

It was noted earlier that Iran has quickly emerged as one of Beijing's key strategic partners for BRI. Bilateral trade with China grew from $4 billion to $13 billion dollars between 2003 and 2013, a trend that is expected to increase significantly as Belt and Road initiatives come into operation. Major urban development, energy, and transport infrastructure projects again form the basis of deals tying Iran's economy into Chinese companies and investors. Just a week after international sanctions on Iran were lifted in January 2016, Xi Jinping traveled to Tehran, with meetings targeting an increase in bilateral trade to $600 billion over ten years. Three months later a new cargo rail service linking the two countries commenced, with the inaugural train arriving in Tehran after the 6,400-mile, two-week trip from Yiwu in eastern China.[22] A proposed $2 billion pipeline will also carry Iranian natural gas into China via Pakistan. Overseen by the China National Petroleum Corporation, the project links up with the company's port and refinery construction program in Gwadar, giving Tehran access to a vast new market in East Asia.[23]

Against this backdrop, Iran has increased its international presence in cultural-sector collaborations. In September 2016, the historical city of Qazvin hosted the Silk Road Mayors Forum meeting, with participation by UNESCO. A month later, cultural exchanges of the ancient Silk Road were the basis of discussions on developing joint tourism strategies during a meeting of the Iran-China Friendship Society in Tehran.[24] During this period, the government of Iran also stepped up its involvement in the Coordinating Committee of the Serial World Heritage Nomination of the Silk Roads, offering to host meetings. The shifting landscape of international relations that Belt and Road is bringing about, coupled with the momentum for listing transboundary Silk Road corridors, could have significant implications for how Iran views its history. As of 2014, Iran had nineteen cultural World Heritage Sites, with a further forty-five on its Tentative List. The 2014 ICOMOS report indicated that forty-six of those—eight listed and thirty-eight tentative—could have their cultural and historical significance reworked to formally tie them in with a more expansive, connective narrative of the Silk Road (figs. 4.5 and 4.6).[25] At the time of writing, it appears as though Iran's Ministry of Culture is looking to more actively engage in these transboundary nominations but has yet to commit particular sites for inclusion on Silk Road corridor nominations. What is clear is that China's

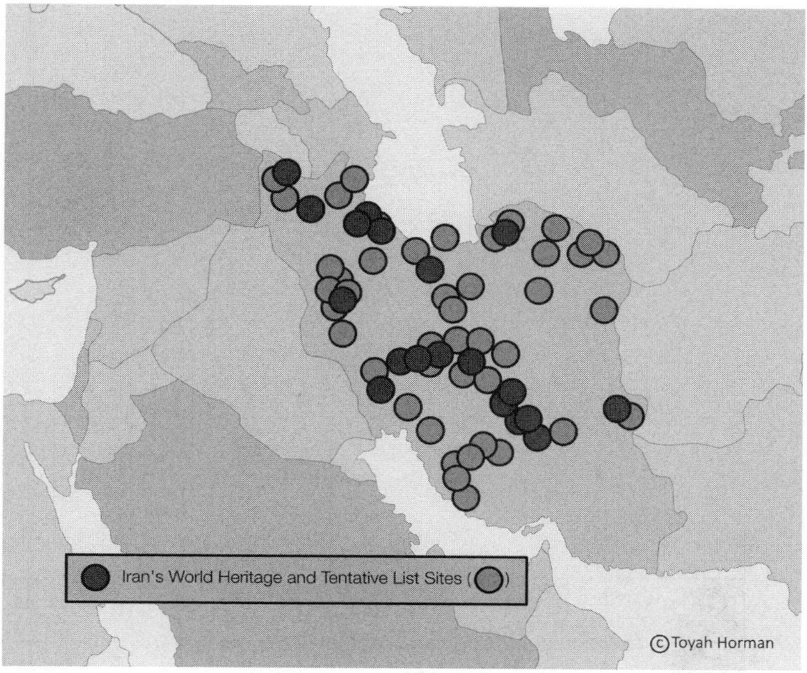

4.5 Iranian sites on World Heritage List and Tentative List. Adapted from Tim Williams, 2014. Courtesy of Toyah Horman.

increased engagement with Central Asia and its need for energy presents significant political and economic openings for Tehran.

Rapid investment in the China-Pakistan Economic Corridor is tying the economies of Iran, Afghanistan, and Pakistan together in new ways. In the case of Afghanistan, the language of the Silk Road offers a whole new narrative for economic development and a counter to Western perceptions of a failing country dependent on foreign aid. The government has responded to the opportunities posed by Belt and Road, emphasizing the country's historical importance as a gateway between the resource rich countries of the Middle East and fast-growing economies of South and East Asia. The opening remarks delivered to delegates to the 2015 Silk Road through Afghanistan conference are reminiscent of S. Frederick Starr's vision for the country, as noted in chapter 2, but more "routed" in history: "Throughout the history, Afghanistan has been a point of intersection in Asia, not a point of deadlock and conflict. The Silk Road was one of the important routes transferring ideas,

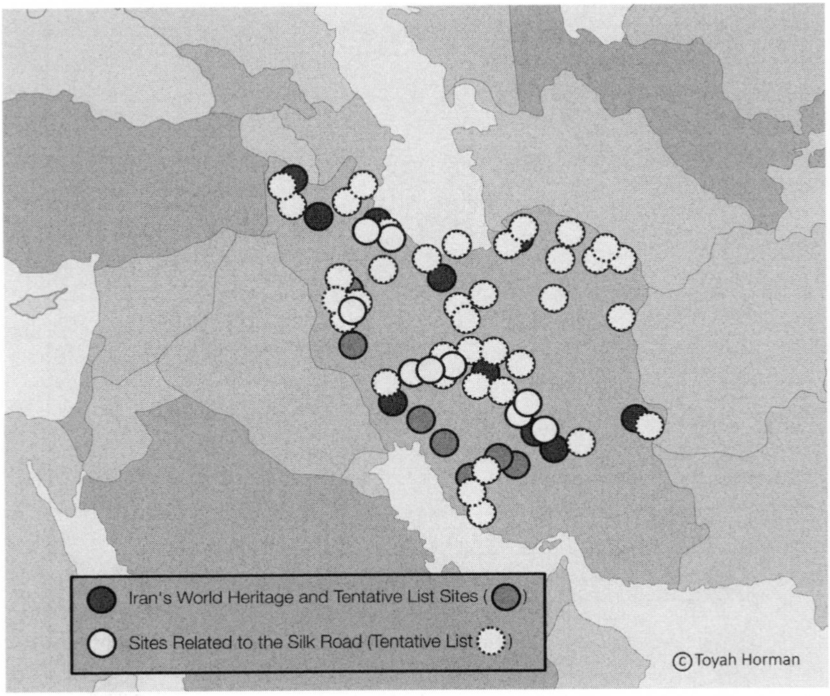

4.6 Proposed sites for listing or renominating as locations along the Silk Road. Courtesy of Toyah Horman.

people and goods across the region that had advanced financial, conflict resolution and comprehensive transportation systems."[26]

Afghanistan, Iran, and other countries in Central Asia are looking to diversify their economies through Belt and Road's emphasis on people-to-people connections. As with inward investments in infrastructure, the likelihood of countries developing Silk Road heritage tourism industries over the medium term varies significantly. Political stability, together with reasonably well-developed tourist infrastructures, mean that countries like Georgia and Kazakhstan are better placed to become mainstream tourism destinations over Afghanistan and Pakistan, which are likely to remain volatile and unsafe for the foreseeable future. Nonetheless, across the region tourism is already regarded as a key mechanism for advancing many of the people-to-people, cross-border collaboration goals of Belt and Road. Over the coming years the Asia-Pacific region will be the key driver in increased tourist arrivals worldwide, expected to climb from around 1 billion per year in 2016 to 1.8 billion by 2030.[27] A rapidly expanding Asian middle class, most notably in China, represents a vast market for developing intraregional tourism oriented around the overland Silk Road. To that end, the World Tourism Organization has brought together representatives from more than thirty governments for a Silk Road tourism initiative that stretches from southeastern Europe to Southeast Asia.[28] Since 2010, they have encouraged visa policy reforms among Silk Road countries to facilitate greater mobility and a less bureaucratic experience for tourists moving across multiple jurisdictions.[29] Xi'an in the province of Shaanxi, often referred to as the gateway to the Silk Road, is among those cities in China seeking to capitalize on this emerging industry. Major heritage-tourism infrastructure projects were quickly slated for development, including the Silk Road International Museum City and Silk Road Expo Park. Initially led by the Tang West Market Group (TWMG) consortium, the proposal pivoted on museums and pavilions exhibiting the arts, crafts, and archaeological artifacts of Silk Road countries from around the region. The group also entered into partnership with UNESCO to create the Silk Road Culture Center in Xi'an, intended to act as a focal point for cultural activities, exchanges, and conferences. In justifying the strategic merits of the project, a report published by the Chinese Ministry of Culture stated: "We have to promote TWMG and other leading cultural enterprises' experience in constructing the Silk Road Economic Belt, bring 'Culture Leads' into full play, and mobilize more social forces to participate in the Silk Road cultural construction and cultural industry development, and provide better service and promote the construction of one belt and one road."[30]

Farther west, Dunhuang, with its famed Mogao Caves, also saw a distinct upswing in cultural-sector activities in the years following Belt and Road's launch. A continuation of the international music forum noted earlier, in 2016 the city hosted the inaugural Silk Road International Cultural Expo. Planned as an annual event to promote heritage tourism and stimulate cooperation across a number of cultural sectors, the expo pulled together performers from seventy countries, with several thousand delegates attending its conferences. As we saw in chapter 2, Dunhuang is a city with strong Buddhist connections to India. *The Hindu* was among the media outlets covering the event for Indian audiences, both flattering the organizers on the "stunning scale" of the event and simultaneously offering a less-than-veiled critique of its purpose: "a strategic exercise of soft-power, to help blunt accusations that Beijing is pursuing self-centred regional dominance, in tune with its economic rise."[31]

In addition to building cross-border ties, the expo is one of a number of initiatives conceived to address a key strategic goal of Belt and Road, that of integrating China's western provinces and its minority groups into the body politic of the nation-state. As we saw in chapter 1, Xinjiang is pivotal to the larger political ambitions of BRI. Building on the "develop the west" strategy starting in 2000, it aims to balance out the geographies of China's urbanization and decades of development in the coastal cities of the southeast through large-scale economic investment in Urumqi and Kashgar.[32] Both cities have been identified as key infrastructure hubs in the development of cross-border trade. Highways starting in Shanghai and central China pass through Urumqi on their way to Kazakhstan, Pakistan, and southwestern Siberia. The city is also a key facilities hub in the Trans-Eurasian Continental Railway, with trains from Beijing routed on to Rotterdam and Madrid. Kashgar is critical to the development of the China-Pakistan Economic Corridor, and it will be the principal point of connection for rail lines currently under construction in Pakistan, Kyrgyzstan, Afghanistan, and Iran, large stretches of which are financed via multibillion-dollar loans from Beijing.[33] Xinjiang's close proximity to Central Asia and histories of migration mean that the region is home to a number of ethnic minorities, including Mongol, Kyrgyz, Tajik, Kazakh, Hui, and Uyghur. The political tensions surrounding the containment of these different cultural groups—most notably the Muslim Uyghur—has received considerable academic scrutiny and cultural heritage has been an arena in which cultural and religious diversity have been celebrated at the same time that difference has been destroyed.[34] For Dru Gladney, BRI success is contingent on a peaceful resolution to the "Xinjiang problem," given that each outbreak of violence undermines the carefully crafted messages projected to the outside world through Belt and Road diplomacy.[35] Moreover, James Leibold cites the warnings of Chinese scholars concerned the country faces the same threat of disintegration as that experienced by the Soviet Union and Yugoslavia, and proceeds to argue that the CCP's desire to maintain harmony between Xinjiang's different minority groups reflects anxieties about a precarious national stability.[36] Today, Beijing is primarily committed to realizing three core goals in Xinjiang, "legitimacy, control and development," and although noting that these are "vital to the consolidation of China's 'rise' as a great power,"[37] Michael Clarke also frames the current situation as the latest iteration in a long history of attempts at managing risk.

In the mid-eighteenth century, the people of Xinjiang endured the expansionist ambitions of the Qing, Russian, and Dzungar Mongol empires. The subsequent consolidation of Qing control, as traced by Millward and others, initially involved the region's segregation and isolation. But as both Millward and James Leibold demonstrate, by the end of the Qing era in the early twentieth century, it was being folded into an emergent national consciousness.[38] Leibold shows how the concepts of frontier and minority emerged as important categories in the creation of the geobody that is China and identification of Chinese nationhood. As we saw in chapter 2, this would also lead to a sense of patrimony over the antiquities recovered from the region in the 1920s and 1930s, despite those decades being characterized by an increase in Xinjiang's autonomy. Claims of ownership by the Nationalist government, the Kuomintang, fluctuated in relation to a complex and violent mix of local rebellions, a conflict with Japan, and the shifting geopolitics of Sino-Soviet relations.[39] The arrival of the People's Liberation Army in Xinjiang in 1949 constituted a takeover by the Chinese communists and the introduction of systematic policies designed to integrate the region with the rest of the country. The deterioration of relations with Moscow also meant that Xinjiang was transformed into a buffer zone, primarily through policies designed to sever economic and cultural ties with the Soviet republics to the West.[40] The People's Republic of China continued an approach borrowed from the Soviet Union regarding the codification of Xinjiang peoples into fourteen nationalities, or *minzu*.[41] This formed part of a nationwide strategy of cultural governance that involved the identification of fifty-six *minzu*. But as Clarke notes, the transition away from Maoism in the late 1970s initiated a change in economic strategy toward Xinjiang. Although its role as a buffer zone was maintained, the region was drawn into Deng Xiaoping's new ideas for China's socioeconomic development. Ongoing domestic reforms, together with the collapse of the Soviet Union, further transformed the political strategy of the PRC, such that a frontier territory began to be reconceived as a strategic bridge to markets and resources across the border in Central and West Asia. Clarke thus argues that this "double opening" not only was about merely establishing new economic ties but also was designed to encourage greater social and cultural connections domestically through increased Han migration, and also to leverage the region's cultural and ethnic ties with the Great Steppe and Islamic Middle East.[42]

The impact of these policy shifts has been the subject of much critical debate in recent years.[43] But with the focus often falling on processes of urbanization, international trade development, or the impacts of Han migration, less critical analysis has been given to how practices of state governance responded to a cultural environment that was changing on the back of emergent cross-border Muslim connectivities. Notable, then, are James Millward's reflections on the Smithsonian Silk Road festival of 2002, discussed in chapter 2. Focusing on the musical traditions of Xinjiang, Millward considered how a Silk Road festival conceived around cross-border trust and dialogue mapped onto a history of Chinese state policy "to preserve, order, develop, and modernize the Uyghur musical tradition."[44] He suggests that Uyghur *muqam*—a musical form rooted in an "Arabo-Irano-Turkic tradition that spans Central Asia, the Caucasus, Afghanistan, Iran, and the Arab countries"—poses particular problems for a PRC nationalist ideology. Accordingly, he stated: "Muqam would thus seem to be the quintessential Silk Road musical form: variations on a theme stretching from the Tarim Basin to the Black Sea. The muqam scales, modes, rhythms, lyrics, instruments and terminology tie the Uyghurs to a system shared across the Islamic heartlands of Eurasia—but do not point to any obvious connections with Chinese musical tradition. Moreover, the fact that the Arab versions maintain consistent modality, while the Uyghur and other Central Asian ones do so only partially or nominally, would seem to indicate an Arab center for the tradition."[45]

Chinese musicologists, he suggested, have contributed to a state effort to de-emphasize such transboundary connections and associations with Islam, instead seeking to ground *muqam* in its local roots and national traditions of Chinese music. Interestingly, and as Millward points out, Uyghurs in Xinjiang have embraced the prestige *muqam* has afforded them, thus imbuing a sense of national pride. Belt and Road brings a new layer of complexity to this situation in that it explicitly invokes a transboundary heritage of cultural routes. Clarke historicizes the second half of the twentieth century as a broad transition from policies oriented around passive containment and isolation, toward Xinjiang's development as a resource of productivity and economic output.[46] Belt and Road appears to take this in a cultural direction; the creation of new markets to stimulate trade in turn stimulates new cultural ties. The promotion of inbound Silk Road tourism would also point to some significant shifts in cultural policy, such that the scenarios of denying the regional historical ties of traditional cultural forms such as *muqam* are replaced with a paradigm of Silk Road connectivity. But the reality on the ground remains more complicated. In 2018 Rachel Harris and James Leibold noted that the politics of intangible cultural heritage in Xinjiang continue to align with policies of surveillance and the forced denunciation of traditional cultural and religious practices.[47] Reports of "reeducation" camps point to a social engineering strategy that is framed as a crackdown on possible sources of violent extremism but that in effect operates as a punitive force of national assimilation against the Uyghurs. Juxtaposing such domestic policies in a region critical to Belt and Road with the CCP's international diplomatic ambitions, Millward highlights the contradictions in play here: "How does the party think that directives banning fasting during Ramadan in Xinjiang, requiring Uighur shops to sell alcohol and prohibiting Muslim parents from giving their children Islamic names will go over with governments and peoples from Pakistan to Turkey? The Chinese government may be calculating that money can buy these states' quiet acceptance. But the thousands of Uighur refugees in Turkey and Syria already complicate China's diplomacy."[48]

Parallel to the analysis of early Soviet-era museology by Francine Hirsch noted in chapter 2, Lu and Varutti have considered the cultural politics at play in the museums of Xinjiang, arguing that representations of heritage and tradition are designed to domesticate difference and to flatten out the sense of hierarchy between different ethnic or religious groups. It is an aesthetic that moves the visitor across the cultural categories, dissolving all political and social relations and any interethnic, minority-majority tensions in a language of diversity. The explicit acknowledgment of cultural diversity through heritage has long been a technology of containment and neutralization across Asia, most notably via "cultural villages."[49] The 2010 Shanghai Expo exemplified this practice. The expo received seventy million visitors over a six-month period—around 99 percent of whom were Chinese. Around 190 countries participated, but the China National Pavilion proved among the most popular, with vast exhibits displaying the traditions, people, cities, and landscapes of each region or province (fig. 4.7). In the area dedicated to Xinjiang, costumed hosts welcomed visitors to multimedia exhibits showcasing the region's food, urban architecture, mosques, and the everyday life of those dwelling in mountainous and desert landscapes. In stark contrast, however, by the time the expo opened its doors, a systematic plan to bulldoze the historic *hutongs* of Kashgar was in its advanced stages. Much of the city's historical cultural heritage was razed to the ground, with the cultural and social ties of communities further ruptured through forced relocations across different parts of the city. Many residents were rehoused in newly constructed high-rise structures.

The key question that remains, then, is whether a discourse of Silk Road connectivity advanced through Belt and Road will create a shift

4.7 The China National Pavilion, Shanghai Expo. Photo by author.

in the cultural politics of minority heritage in Xinjiang over the longer term. Belt and Road suggests the Islamic heritage of cities like Kashgar now have an altogether-different political expediency, affording new cultural connections with neighboring countries and populations, the people-to-people connections of strategy documents. World Heritage listings invariably lead to management plans that affect the lives of local residents and infrastructure schemes that transform rural and urban spaces alike. Parallel to the "opening up" of Tibet in the early 2000s, it is likely that the prospect of domestic and cross-border tourism and economic development will increase the number of Han Chinese entrepreneurs seeking opportunities across Xinjiang. World heritage will also serve as the catalyst for new museum projects and policies and legislation for intangible cultural heritage. Kashgar has hosted a small and largely neglected Silk Road museum since 1992 featuring ceramics, paintings, rugs, calligraphy, and handicrafts of Persian, Turkic, and Indian origin. The idea of a historical Silk Road places the city and the other parts of Xinjiang firmly in the center of the cultural politics of Belt and Road. On the one hand, BRI suggests a future of greater connectivity, whereby previously denied cultural and religious ties are reactivated and even proactively supported as part of schemes to develop Xinjiang's cities as gateways for development corridors heading out west and south. On the other hand, it is highly likely to continue previous policies of containment. This prompts important questions concerning how the paraphernalia of cultural heritage—museums, tourist sites, urban conservation projects—will come to act as technologies of surveillance or cultural repression, or whether Northwest China will be formally recognized by the Chinese state as a contact zone of religions, culture, and civilizations in ways that alleviate some of the threats to minority identity, language, and ways of life that have been in place for decades.

Heritage Corridors by the Sea

At sea, the Initiative will focus on jointly building smooth, secure and efficient transport routes connecting major sea ports along the Belt and Road.

VISION AND ACTIONS STATEMENT, MARCH 2015[50]

If we return to the Maritime Silk Road, the notion of trade or economic corridors becomes a little more diffuse, as the infrastructure required is not the connective tissue of pipelines, railways, or highways. Nonetheless, what we are seeing in the Maritime Silk Road is the establishment of discernible routes for commercial shipping, made viable by a network of ports, special economic zones, and other coastal facilities. Maritime histories have once again played a role here in ways that predate Belt and Road, as we saw in the previous chapter. My interest in the pages that follow is in revealing the broader trajectories this process has followed over the past twenty years or so, pointing to evidence of a growing use of heritage diplomacy across the seas and the increasing prominence of Zheng He. I will suggest that the Maritime Silk Road of the twenty-first century takes this to a new level, introducing new players and a new narrative for injecting funds into archaeology, conservation, museums, and heritage tourism attractions.

China's first forays into state-funded maritime archaeology began in November 1987, with the establishment of the Underwater Archaeological Research Centre (UWARC), located within the Ministry of Chinese History. In 1989 new legislation was passed regulating underwater archaeology in territorial waters, with a specific prohibition on any unlicensed commercial salvage operations. Jeff Adams argues that the operations of UWARC have been closely tied to China's territorial interests in the South China Sea and beyond. Accordingly, he cites UWARC's first open-water project in the mid-1990s, which took place in the Paracel Islands, a location that has been the focal point of contested ownership claims between China and Vietnam since 1974.[51] Wrecks such as the Song dynasty *Huaguang Reef 1* were salvaged to give weight to claims of sovereignty based on historical grounds. As Adams explains, public pronouncements to this end were officially refuted by Vietnam's embassy in Beijing.[52] At a broader level, maritime archaeology has also advanced a narrative of redemptive civilizational grandeur; the idea of China as a great power that secured its wealth and respect via ocean-based commerce and peaceful diplomacy. The excavation of the *Nanhai-1* from the Pearl River delta proved instrumental in this regard. Discovered in 1987 but not salvaged until 2001, the 102-foot Song dynasty *Nanhai-1* is larger than the much-celebrated wrecks of the *Vasa* and *Mary Rose* on display in Stockholm and Portsmouth, England, respectively. Adams nicely captures the significance the wreck and its excavation holds, stating that the *Nanhai-1* "perfectly unites notions of past enterprise, present achievement, and national destiny."[53] Not surprisingly, an elaborate museum designed to showcase the wreck and three hundred thousand of its artifacts was established in the coastal city of Yangjiang. Since its opening in 2009, this Maritime Silk Road Museum of Guangdong has added a second exhibition, *The Maritime Silk Road*, dedicated to China's maritime connections with far-flung Buddhist, Christian, and Muslim communities.

The resources invested in the *Nanhai-1* form part of a larger state-funded program for maritime archaeology, which has gathered momentum over the past two decades. Laws have been revised to bring the country in line with UNESCO's convention on underwater archaeology, and a number of research centers and museums have opened in an explicit attempt to raise public consciousness about China's maritime past. In 2009 the National Conservation Centre for Underwater Cultural Heritage was founded with the remit of coordinating cultural heritage research and management across a number of provinces. Under this program, research centers and museums were created in Ningbo, Wuhan, Fujian, and Qingdao.[54] Together these have greatly increased efforts to find and salvage wrecks in the East and South China Seas. Since the early 1990s, around a dozen significant shipwrecks have been excavated.[55] In 2010 China joined a select group of countries with crewed submersibles capable of operating below depths of three thousand feet. These have been used to conduct surveys of the seabed in deeper waters beyond the continental shelf. Four years later, the *Kaogu-01* was launched, China's first vessel designed specifically for underwater archaeology, complete with folding cranes, workrooms, and air lock chambers for cultural relics recovered from the seabed. In the past few years, new surveys have been conducted in the disputed Spratly and Paracel Islands. Interestingly, while a few astute observers have linked such activities to the larger political tensions of the South China Sea, maritime archaeology has received far less attention than the construction of artificial islands. And yet history and submerged cultural resources have long been a central pillar of China's claims over the waters. In the run-up to the decision of the Permanent Court of Arbitration at The Hague, a report submitted to the court asserted ownership over the Spratly and Paracel islands on the basis that China was the first to discover and name the island groups: Nansha and Xisha, respectively. For Adams, then, there has been a clear set of intersections as archaeological expeditions and their technologies have become increasingly entangled in strategies to monitor and surveil disputed waters, and the policies of government ministries engaged in marine industry and military affairs.[56] Indeed, I would argue that we are seeing a convergence between the search for underwater antiquities and the construction of infrastructure within a geopolitically contested space in ways that mirror the Great Game of Central Asia in the late nineteenth century.

Not surprisingly, the figure of Zheng He has been an ever-present feature of this story. UWARC has implemented a number of searches for vessels from Zheng He's fleets. Elsewhere, scholars have sought evidence to document the extent and nature of his treasure-ship voyages. The vast majority of the records stored in archives in Beijing and Nanjing were burned during fighting at the end of the Ming dynasty. Those nautical charts and documents that did survive, together with Chinese ceramics found in Southeast Asia and along the East African coast, have underpinned scholarship on Zheng He within China, which emerged in the 1980s.[57] The motivation for this research has invariably come from a desire to see him given greater recognition within the narrative of China's past as well as that of global maritime navigation.[58] The six hundredth anniversary of Zheng He's first voyages celebrated in 2005 offered an opportunity for taking this research into a wider public domain. Museum exhibitions were opened, statues were erected, and replicas of Ming dynasty vessels were built in several cities. In Nanjing, the newly renamed and excavated Zheng He Memorial Shipyard hosted extensive celebrations over the course of the year. In addition to a treasure-ship replica rebuilt with a budget of $10 million, the revamped shipyard included corridors of murals depicting evocative scenes from his encounters with foreign lands, and their exotic

4.8 Replica treasure ship, Zheng He Memorial Shipyard, Nanjing. Photo by author.

people and animals (figs. 4.8 and 4.9).[59] In 2008, the seven voyages of Zheng He featured prominently in the opening ceremony of the Beijing Olympics, with hundreds of performers simulating the rowing action of his treasure ships with decorated oars. Two years later, a full-size Ming dynasty replica served as the centerpiece for the newly opened China Maritime Museum in Shanghai. This canonization of Zheng He within a language of maritime heritage has emerged in a rapid and highly choreographed manner, whereby he has become a symbol for maritime power and for a nation that looks outward to build its wealth and influence in the world. As with all discourses of cultural heritage, it thus speaks to the anxieties and ambitions of the present more than the histories it selectively represents.

As we saw in the speeches from the previous chapter, this political appropriation of Zheng He has extended far beyond the borders of China itself. Indeed, through different forms of international collaboration, he has increasingly emerged as part of the heritage industries of Southeast Asia and East Africa. Across the region, 2005 proved a catalyst in the popular memory of Zheng He. Extensive celebrations were held to commemorate the six hundredth anniversary of his voyages, with festivals and exhibitions staged in Indonesia, Malaysia, Singapore, and the Philippines. Given that no records show Zheng He's fleets ever visited Singapore, the evolution of a maritime heritage predating British settlement that attempts to maintain the city-state's dominant position in twenty-first-century commercial shipping is particularly revealing, as well as a story picked up in greater detail in the following chapter.

Beyond Southeast Asia, among the various other countries that celebrated Zheng He's maiden voyage, two in particular, Sri Lanka and Kenya, would subsequently dig deeper and literally excavate their coastlines in the hope of finding evidence of landings and catastrophic sinkings—coinage in the currencies of diplomacy, old and new. In Kenya records indicate that two ships were lost near the historical town of Lamu. On the back of stories claiming survivors swam to shore, Chinese scientists traveled to the Kenyan island of Pate to conduct DNA tests on residents in an attempt to establish their Chinese descent. In what became an extraordinary story, the young Kenyan woman Mwamaka Sharifu gained notoriety for taking up a Chinese government scholarship to study traditional medicine in Nanjing. Widely reported in China as having Chinese blood and given the sobriquet "the China girl," Sharifu became a doctor upon graduation and remained in China to play an informal ambassadorial role between the two countries. Around the time of Belt and Road's launch, efforts to find evidence of Zheng He's

4.9 Murals depicting voyages of Zheng He, Zheng He Memorial Shipyard, Nanjing. Photo by author.

lost ships were stepped up, with the Chinese government committing $2.4 million in maritime archaeology assistance.[60] Conducted as a collaboration between Peking University and National Museums of Kenya, the project was justified through the discovery of porcelain artifacts containing dragon symbols. Despite not finding a shipwreck, in late 2016 Kenyan news outlets continued to report that Ming-era porcelain and coins were being discovered off the coast and in villages.[61] For the Washington-based archaeologist Chapurukha Kusimba, such finds were highly significant for Kenya: "Zheng He was, in many ways, the Christopher Columbus of China. . . . It's wonderful to have a coin that may ultimately prove he came to Kenya."[62] Not surprisingly, within the country itself there was much enthusiasm for any evidence of longstanding connections with China. Lamu's Swahili history museum was among those linking shards of pottery to the voyages of Zheng He. In late 2013, it was also reported that the National Museums of Kenya intended to build the country's first underwater museum. The culmination of a three-year partnership with Chinese archaeologists, the project was designed to display the artifacts found by the team. After some initial delays, construction on the museum commenced in 2015.

All this has taken place in a context of rising Chinese investment in Kenya and the country's strategic importance as a gateway to East Africa. Over the past decade or so, Nairobi and Mombasa have become key infrastructure nodes, with multibillion-dollar investments going into port, pipeline, rail, and electricity production projects.[63] On the back of rapid growth in bilateral trade between the two countries, climbing over $5 billion in 2014, Lamu is set to become a catalyst for the development of the East African region under Belt and Road.[64] The construction of a vast new "megaport" in Lamu—a project launched with a $480 million funding pack from the China Communications Construction Company—is the first step toward a project valued at $23 billion, which, if fully completed, will remap the trade and infrastructure networks of Kenya and its neighboring countries via a Sudan-Ethiopia transport corridor.[65] Built over several phases, the Lamu port is expected to eventually consist of thirty berths, at an estimated cost of around $3.5 billion.[66] In May 2014, signatures were also placed on a $3.6 billion contract to connect Mombasa to Nairobi by rail, the financing for which would be 90 percent covered by China.[67]

These intersections of the past and present Maritime Silk Roads have also defined Sri Lanka's recent engagements with China, as we saw in the previous chapter. Paralleling Kenya, Sri Lanka also marked the sexcentenary of Zheng He's voyages and his six visits to the island. After some years of discussion, two controversial maritime archaeology agreements designed to expedite the search for Zheng He wreckage off the Sri Lankan coast were signed in 2012 and 2014. The first focused on the waters of Hambantota, the location of a Chinese-funded deepwater port. The second involved surveying waters off the historical port city of Galle. As a collaboration covering both museums and underwater archaeology, the contract stipulated that a percentage of the recovered antiquities would travel to China.[68] These projects have been part of bilateral agreements involving maritime and coastal security, free trade, tourism development, and infrastructure projects.[69] Central to the discourse of cooperation has been the idea that both countries share centuries of trade and cultural exchange. In Sri Lanka's plans to develop its tourism industry, increasing attention has been paid to the material evidence of Zheng He's visits to the island. The prominence given to a stone tablet inscribed with texts in Persian, Chinese, and Tamil, the three primary languages of trade at that time, represents one such example. Erected in 1409 to mark Zheng He's second visit, the tablet was discovered by the Public Works Department in 1911 in Galle, and eventually made its way to the National Museum of Sri Lanka, Colombo, where it remains on display today. In May 2014 interpretative texts in Chinese were added and four months later Mahinda Rajapaksa presented Xi Jinping with a rubbing of the tablet's inscription to mark the occasion of the first visit to the island by a Chinese head of state in three decades. Two further replicas are located in the Maritime Museum in Galle and the Treasure Boat Shipyard Park in Nanjing. An exhibition hall dedicated to Zheng He has also been constructed in the Galle museum.

As does Kenya, Sri Lanka plays a critical role in the development of oceanic trade for China. As part of Beijing's strategy of building deepwater port facilities around the Indian Ocean—the so-called string-of-pearls program—projects in Colombo and Hambantota have been funded through loans of $1.4 billion and $1.7 billion, respectively. In addition to a newly built commercial port, the Colombo Port City Project involves the construction of shopping malls, residential towers, a marina, and tourism infrastructure. The project stretches across more than 550 acres, 217 of which have been leased to the China Communications Construction Company for ninety-nine years.[70] Further deals were signed during Xi Jinping's visit in September 2014. Nestled among the launch of the final phase of a $1.3 billion, nine-hundred-megawatt power plant, and talk of developing the Maritime Silk Road for the twenty-first century was an expressed concern to reestablish ties of friendship dating back hundreds of years.[71] The power plant, together with earlier port and highway projects, form part of an overall Chinese investment program in Sri Lanka, estimated in 2015 to be worth in excess of $5 billion.[72] The coming into office of President Maithripala Sirisena in early 2015 suggested that Sri Lanka's foreign policy was taking a change in direction, most notably via a nuclear energy deal signed with India a month after the election. A year later, however, the government renewed its commitment to closer ties with Beijing and its role in Belt and Road. At the eighth BRICS summit, held in Goa in October 2016, both presidents agreed that the sixtieth anniversary of diplomatic relations between the two countries in 2017 offered an opportunity to strengthen ties across tourism, security, disaster preparation, media, and to build exchanges between local governments, youth, and the two countries' Buddhist communities.[73]

Kenya and Sri Lanka are two examples of the promotion of a Zheng He heritage industry being incorporated into the bilateral agreements of central government. Elsewhere, however, his profile has been raised at a more local or regional level. But as the examples from Malaysia and Indonesia that follow illustrate, the advent of Belt and Road meant that these historical connections with China were once again co-opted by the state. As the Straits of Melaka became vital to fifteenth-century international trade, the port state of Melaka represented a strategically valuable staging post for Zheng He, with records indicating five of the seven fleets stopping off there. As Geoff Wade suggests, "The links between Melaka and the Ming thereby remained intimate for much of the first half of the fifteenth century."[74] Storage depots, or *guanchang*, were constructed in the port. Depicted in maps from the time, these structures have been a point of contention in the debate about the nature and intentions of the fleets and voyages led by Zheng He, or Cheng Ho, as he is more commonly known in Southeast Asia. Wade describes them as "military garrisons-cum-treasuries."[75] In contrast, Tan Ta Sen, owner of the Cheng Ho Cultural Museum in Melaka, maintains that they were used for commercial purposes only, primarily for the storage of currency and food.[76] Tan also claims that the timber building housing the museum alongside the Melaka River is located on the site of an original *guanchang*, citing evidence of recovered artifacts and a well.[77]

The Melaka museum represents one arm of a long-term effort by a group of historians and businesspeople to promote the memory of Cheng Ho in Southeast Asia. In 2002 Tan founded the International Zheng He Society, becoming its president a year later. With an advisory board made up of diplomats and senior academics from Indonesia,

Malaysia, and Singapore, the society has raised awareness about Zheng He's voyages and promoted the study of their impact on Southeast Asia. As Tan explains "to the members of the International Zheng He Society, Cheng Ho was an envoy of peace, who played an important role in the diplomatic, cultural, and economic exchanges between China and foreign countries in the 15th century. He stands today as a symbol of equality and racial harmony."[78] The impact of the society on Melaka's heritage-tourism industry has been tangible. A replica of a Ming dynasty ship is the centerpiece of a key traffic intersection in the old town, billboards promote Cheng Ho, and cafés, restaurants, and boutique hotels all take their name from his voyages (fig. 4.10). This positive endorsement of Zheng He aligns with Malaysia's ambitions to be a key player in the China-Indochina Economic Corridor of today, with much of the investment focusing on Melaka itself. Once completed, the Melaka Gateway Project will be the largest port in the region. By making multibillion-dollar investments in a number of Malaysia's state-owned businesses, China has gained a foothold in the country's energy, real estate, construction, and finance sectors. Given the strategic

4.10 Model of ship from Zheng He fleet, Melaka. Photo by author.

importance of Melaka and its port on the Maritime Silk Road, Beijing committed up to $10 billion over ten years, starting from 2015.[79]

In Indonesia, the legacy of Zheng He has been somewhat different. Rather than being revered as a historical figure, his canonization has taken place through Islam. Under the New Order regime of President Suharto, hundreds of thousands of ethnic Chinese endured decades of persecution and violence, with his pro-US government casting them as scapegoats for poor economic performance and the specter of communism. As conditions changed in the years following Suharto's 1998 resignation, legislative changes were made to enable greater political and social integration of ethnic Chinese communities across the country. This period, known as Reformasi, created a new public discourse of Chinese identity, with a more pluralist, democratic political environment enabling a revival of Chinese language, culture, and religion. Johanes Herlijanto describes the role Zheng He has played in this change, noting that the religious aspects of his voyages have been given particular prominence alongside the now-familiar themes of friendship and mutual cultural understanding.[80] Quoting from interviews with academics, journalists, and politicians, Herlijanto demonstrates how the idea that Zheng He brought Islam to the islands has been successfully propagated by a number of Chinese Indonesian groups over the past fifteen years or so.[81] Public festivals, most notably celebrations for the six hundredth anniversary, together with theater and other cultural performances, and even a television sitcom, have all contributed to a wider acceptance of China and the role Chinese Muslims have played in Islamizing Indonesia. Within this reframing of the past for the political present, religion and different architectural forms have been fused together to create a new language of transnational cultural heritage.

In Semarang, Java, for example, the fifteenth-century Chinese temple Sam Poo Kong, which incidentally takes its name from the popular term for Zheng He, is today shared by multiple religious denominations. Values of tolerance and openness are the message of the various shrines, statues, and mural displays celebrating his visit to the island.[82] Farther east, the newly built Cheng Ho Mosque was inaugurated in Surabaya in 2002. Designed to accommodate two hundred worshipers, the mosque took inspiration from Beijing's much-larger Niujie Mosque, which began welcoming worshipers into its doorless buildings in the Islamic and Han Chinese style in 1996. On the island of Sumatra, the city of Palembang opened its new Muhammad Cheng Ho Mosque in 2008. This time the architecture drew on Malay and Chinese motifs, with two large minarets mimicking Chinese pagodas dominating the site. In considering the wider significance of Zheng He in Indonesia today, Herlijanto argues that his story carries two distinct messages. On the one hand, discourses of tolerance and multiculturalism are productive in countering the discrimination and conflicts that continue to arise from Indonesian ethnic nationalism. On the other hand, Zheng He embodies the idea that China brings peace and prosperity to the country. Suspicions are growing in Indonesia concerning China's sizable investments in the country and how those investments benefit the ethnic Chinese business elite. Reports of the China Development Bank funding the Bandung-Jakarta high-speed rail project and an ambitious proposal to build seventeen islands in the Bay of Jakarta for real estate and flood prevention are evidence of deepening ties between Jakarta and Beijing. Indeed, the launch of the 21st Century Maritime Silk Route Economic Belt at the Indonesian parliament in October 2013 further strengthened this cooperation. The following year China's total direct investment into Indonesia grew an astounding 37 percent, to just over $3.8 billion.[83] Beijing's embassy in Jakarta proclaimed a future of coprosperity as bilateral trade, calculated to be worth $63 billion, continued to grow, in large part through the Chinese businesses—a number they estimated to be in excess of one thousand—which had begun operating across all sectors of the Indonesian economy. In 2015, China's Vice Premier Liu Yandong visited Jakarta again, declaring her government's desire to strengthen the people-to-people exchanges between the two countries.[84] Within Indonesia, however, this has led to considerable unease about the growing influence of both China and Chinese Indonesians on the politics and on the economic and social life of the country. However, for those invested in seeing stronger ties between China and Indonesia through Belt and Road, Zheng He continues to represent a unique form of heritage diplomacy. Choirul Mahfud, for example, argues that Surabaya's Cheng Ho Mosque links the two countries in cordial relations, past and present. Accordingly, he points to cleric exchange programs, the hosting of Chinese cultural events, publications, and the promotion of Islam in China as activities that the government should support as part of its strategy for building the trust and networks required to "revive" the Silk Road.[85] In 2014 the Indonesian government introduced a Zheng He–themed tourism program to encourage Chinese tourists to the country. A year later, this was incorporated into a larger bilateral agreement signed at the 2015 Boao Forum, which included inbound tourist goals of two million on both sides and a series of agreements "to promote cooperation in cultural heritage tourism."[86]

In the various examples outlined here, the discourse and propaganda of Zheng He has been consistent: that of friendship, trade, and exchange. Up until 2013 and the launch of the Maritime Silk Road this was a diplomatic narrative deployed primarily to counter negative perceptions of China and anxieties over its growing economic and military presence across the region. Ostensibly, the South China Sea has been a territorial dispute. Beijing's proposal to rebuild historical trade routes through the region shifts Zheng He from being a counter-defensive strategy to an asset for the vision of an outward-looking, mercantile China of the twenty-first century. In other words, once framed within the concept of a Maritime Silk Road, he aligns the present with the past, the historical forefather of a grand strategy for a revival of "peaceful trade." In this respect, as the narratives of Zheng He and the policies of the Maritime Silk Road are locked together, they deterritorialize the seas of East and Southeast Asia and create new assemblages of regional connectivity, flow, and mobility. Beijing's participation in the Year of China-ASEAN Maritime Cooperation is indicative in this regard. Speaking at the program's 2015 launch event, Yang Jiechi, a prominent architect of China's foreign policy, invoked a premodern era of cosmopolitan trade, one that predates territorial claims and national disputes over fishing and energy rights, to harness future intergovernmental dialogue:

> The ancient maritime Silk Road was opened and operated by the people of Asia, Europe and Africa. For hundreds of years, a large number of ports thrived along the coastlines on the West Pacific, the South China Sea and the Indian Ocean. Calling on these ports were not only ships carrying silk, but also those loaded with porcelain, ironware, spices, precious stones and books. Crew and passengers included Chinese, Europeans, people from Southeast and South Asia, Arabs and Africans, who came to each other's aid readily as travellers in the same boat. In the early 15th Century, the great Chinese navigator Zheng He who led the world's biggest fleet at the time on a total of seven expeditions, reaching as far as East Africa, the Red Sea and the Persian Gulf. They did not invade, colonize or swindle, but went for trade along with spreading amity and cracking down on piracy. Zheng He's fleet received welcome and assistance from the countries along the route and touching stories about Zheng He are still being told to this day.
>
> In 2013, President Xi Jinping put forward the major "Belt and Road" initiative. Countries and people along the route have applauded it. We all agree to make the building of a community of common destiny for win-win cooperation the goal of our endeavour. And we all agree to move forward the building of overland and maritime silk roads in tandem. Building the 21st Century Maritime Silk Road is the

continuation and development of the ancient maritime Silk Road. What we want to continue pushing forward is the Silk Road spirit featuring peace, friendship, openness, inclusiveness, mutual learning and mutual benefit.[87]

It is a change in emphasis to which others can connect. As we have seen, those countries wishing to be part of the Silk Roads of the twenty-first century find strategic merit in previous histories of contact and commerce. Of course, this celebration and commemoration of past connectivity is not a determinant for where China selects its infrastructure investments for transnational trade. But as the mythology and manipulation of Zheng He, together with the idea of a historical Maritime Silk Road, have gained traction, maritime history—its wrecks, admirals, port cities, cosmopolitanism, and material artifacts—has emerged as an enabler of government-government relations, international investment projects, and the people-to-people connections (to use the language of Belt and Road) associated with tourism. Overseas Chinese tourism to Sri Lanka and countries across Southeast Asia has dramatically increased in the past decade, in many cases through the activities of Chinese business communities. Zheng He and the evidence of Chinese cultural influence has become a popular marketing strategy for the outbound tourism industry in China. In Taicang, Jiangsu Province, for example, the voyages of Zheng He were the focus of a tourism promotion program to Sri Lanka.[88] This has created new discourses of heritage in South and Southeast Asia. Overseas Chinese communities and their host governments have capitalized accordingly, promoting attractions, foods, souvenirs, and experiences that appeal to the Chinese tourist. The following chapter further examines such themes, identifying how the broad narratives of connectivity and exchange are materialized in different ways. Examples from Singapore and the Bay of Bengal also show how cities and governments are deploying material culture in trying to insert themselves into the matrix of Silk Road connectivity.

Clearly, the rhetoric of the ancient Silk Roads glosses over extraordinary complexities in the domestic and international politics of Eurasia. In all the proclamations cited in these two chapters, there is an explicit and deliberate avoidance of conflicts of any form, whether at the domestic or the international level, historical or contemporary. But Belt and Road is faced with very real political and geographical challenges. In Central Asia, the vast Fergana Valley—straddling the post-Soviet states of Kyrgyzstan, Uzbekistan, and Tajikistan—is home to a diverse array of groups whose ethno-nationalist histories remain a source of potential instability within and across national borders.[89] In Afghanistan and Pakistan tribal factions and militant movements are likely to define the contours of domestic politics for many years to come, as are the deep-seated hostilities between Shiite and Sunni Muslims across the Middle East more broadly. It would be foolish to assume that such issues will not bear heavily on the future of BRI, undermining and destabilizing the plans of business and political leaders across the region. We have also seen how domestic heritage issues bear on international diplomatic agendas. Indeed, the economic and development corridors of Belt and Road will no doubt stutter and stumble, and in some cases fail to materialize in the ways that agreements and master plans envisage. The examples given here represent, then, a combination of partially implemented and completed projects, as well as a series of aspirations which have varying degrees of viability over the medium to long term. The example of Xinjiang also highlights how states are drawn into various risk calculations. Domestic priorities of stability and harmony mean that Belt and Road could uphold, even accelerate, punitive cultural policies toward minority groups. Four decades of World Heritage in Asia have written a story of forcibly displaced communities, the elite capture of resources and profits, and rising prices for local residents. I contend that the future development of Silk Road World Heritage corridors will further such processes, and, following Millward's observations, on-the-ground developments will undoubtedly bear on the international ambitions of governments seeking to capitalize on Belt and Road. Over the longer term, Silk Road pasts may well provide new politically productive connections and opportunities for various social groups, including those living in Xinjiang. Equally, however, it is likely that states will co-opt a history of transboundary routes and connections to reinforce existing policies.

In looking to the intersections of heritage-related projects and the other elements of Belt and Road, the aim is not to suggest a series of causal relations, whereby the cultural directly leads to multibillion-dollar port agreements. Rather, it is about highlighting the diplomatic affordances of history and heritage, about how they provide the symbolic and discursive framing around which relationships are woven together. The various examples here have pointed to the ambitions of building transboundary corridors, where connectivity is politically and financially engineered into existence along certain routes. As we have seen, Belt and Road builds on certain preexisting connections and networks, multiplying their effect and expanding the number of nodes. Melaka challenges the hegemony of Singapore in the Southeast Asia container-shipping industry, and refineries in Myanmar and Bangladesh fundamentally alter energy-supply routes to East Asia. In what is a fast-changing and increasingly networked regional economy, countries and cities are thus looking to secure advantage and visibility. Having traced some of the ways in which archaeological sites, communities, and remote Silk Road cities have been drawn into this twenty-first-century world of deal making, the following chapter brings the analysis down to the scale of the object to highlight how stones and timber, silk and porcelain, are strategically narrated in an era of unprecedented connectivity.

FIVE

Objects of Itinerancy

In the past twenty years or so, interest in industrial heritage around the world has, in many cases, cast its gaze toward infrastructures of connectivity. Bridges, canals, and railways are among the most celebrated legacies of what was a new age of mobility and carriage. In England fathers even name their sons after the industrial engineering achievements of national heroes, as I can testify, having been given the appellation of a suspension bridge designed by Isambard Kingdom Brunel in Clifton, Bristol. But unlike the industrial heritage of the nineteenth century, wherein railways, bridges, and other infrastructure are now being preserved and restored, the connective tissue of the Silk Roads, the actual material fabric that the sector so often relies on to tell its stories, has long since disappeared. For a history in which camel caravans and sailing ships carried goods across great distances, the strongest forms of evidence lie on the seabed, and perhaps buried in sand dunes, awaiting recovery. Instead, the story of the Silk Roads comes alive in the landscape through its "nodes" of connection, such as coastal ports and market towns. But given that the Silk Roads are celebrated as a story of carriage, mobility, and exchange, it is also helpful to turn to another material history, that of the objects of itinerancy. As noted in chapter 1, a vast array of goods were carried across land and sea over the centuries. Beyond the famed raw and finished silks, teas, spices, gunpowder, gems, ivory, and furs, there were glassware, furniture, navigation technologies, and exotic animals all packed for shipment. Today, those items that have survived the centuries impart invaluable knowledge to our understanding of history, and those objects have become key actors in the heritage discourses for the two Silk Roads.

In this regard, the Silk Roads offer ample opportunities for writing the style of commodity-chain histories that have attracted wide audiences in recent times. On screen and on paper, foodstuffs have become particularly popular, with salt, coffee, sugar, and spices among the commodities through which stories of exploration, slavery, or imperial trade have unfolded. In the case of the Silk Roads, certain objects have also emerged as metonyms for the overarching narratives of exchange, dialogue, and encounter. But as material remnants are drawn into the political relations of today, their significance is amplified, and they begin to circulate in altogether-different regimes of value. What I hope to show, then, is that they also figure into the crafting of diplomatic relations. The aim is to highlight how certain objects and their itinerant histories afford new forms of international relations and diplomacy today. Once again, the following pages highlight a series of trends set in motion before the launch of Belt and Road that have continued to find new political and diplomatic resonances and connections since 2013.

Over time a select group of objects have been institutionalized as metonyms of the Silk Road, symbolically and physically encased for public display. I limit my focus to some illustrative examples, namely porcelain and some technologies of its carriage. Compared to silk, ceramics are significantly more durable. In this respect, they open up some interesting possibilities. Broadly speaking, the term *Silk Road* has been synonymous with the overland route. But as investments are made in finding, preserving, and displaying objects of travel, including the tens of thousands of ceramics already recovered from the seabed, the complex and long histories of maritime Asia are likely to become far better known. The themes explored here help illustrate the merits of using an object-centered history as an organizing frame for understanding Silk Road heritage diplomacy today. The chapter considers how artifacts form part of emergent heritage assemblages conceived and oriented around histories of transnational mobility, as cities and governments hope to put themselves on the map of Silk Road connectivities, past and present. The final section of the chapter looks forward by anticipating some possible futures for objects that were not necessarily part of the story of Silk Road itinerancy but have been violently relocated in recent times through conflict looting. It is suggested that as the different cultural, political, and physical elements of Belt and Road come together, former Silk Road cities in the Middle East—Palmyra, Aleppo, Bosra, and Damascus in Syria, or Hatra, Erbil, and Mosul in Iraq—could become the source of new forms of trafficking, as museums and private collectors in East Asia and elsewhere in the region seek to build their collections.

It is a line of inquiry drawing on recent thinking that frames the value of objects in terms of their social lives and mobilities. Arjun Appadurai's notion of scapes, together with Igor Kopytoff's cultural biography of things and James Clifford's traveling cultures, provided the foundations for analyses of material culture built around ideas of pathways, routes, and life spans and a metaphorically conceptual language of biography and travel.[1] Of particular importance is the way in which this has shed light on how objects attain value through their movement between social settings and in movement itself. We now better understand how cultural value arises through modes of transmission and through exchanges and relations. As Hahn and Weiss remind us, for example, hierarchies of value can emerge in a particular social setting as objects from elsewhere are assigned greater value.[2] In an attempt to offer a more precise depiction of what is at stake in movement and the transpositions that occur between contexts, I share their preference for the term *itinerary* over *biography* or *travel*. As they suggest, it helps to not just "emphasise a mobile form of existence," but also points our attention to the ways objects travel along previously identified routes and come to rest at stopping-off points along the way.[3] As objects follow such itineraries, they are propelled by social and economic processes. The spotlight here falls on some specific examples, namely ships, ceramics, and the movable heritage of Silk Road cities torn apart by war. Considerable attention is given to the importance of ceramics in the material culture histories of Belt and Road diplomacy. As we will see, their durability, portability, and designed forms reveal not just histories of trade networks but also the routes of aesthetic appreciation and cultural transmission. Other objects could have been chosen. Michael Alram argues the case for reading Silk Road histories through coins.[4] And in 2017, the Museum of Islamic Arts in Doha provided a fascinating example of this turn toward connectivity through its *Imperial Threads* exhibition. Carpets, decorated manuscripts, and ceramics provided the material evidence of artistic exchange across Central and West Asia, specifically between Safavid Iran, Mughal India, and the Ottoman Empire (fig. 5.1). The argument here is that Belt and Road has advanced this trend in discernible ways and will incorporate further objects as a heritage of itinerancy continues to gain political import.

5.1 *Imperial Threads* exhibition, Museum of Islamic Arts, Doha. Photo by author.

Shipments of Porcelain

A small Arabian dhow, just under sixty-six feet in length, discovered just off the Indonesian coast in 1998 is now regarded as one of the great archaeological finds of modern times. Stumbled upon by fisherman in fifty feet of water just off the island of Belitung, the vessel was laden with cargo bound for the Middle East. Archaeologists are confident that the Belitung, as it has come to be known, sailed from Guangzhou in southern China, but it remains unclear why it ventured some distance south from its expected turning point for passage through the Straits of Melaka. It is believed the dhow was built along the coast of either the Persian Gulf or northern India and that its final, fateful voyage took place at some point around 830 CE.[5] Its likely destination was either Oman or Basra in Iraq.[6] Since its discovery, the Belitung has been the subject of considerable controversy, with the salvage operation undertaken by Maritime Explorations, a private firm operating under license from the Indonesian government. This situation contravened UNESCO's guidelines for underwater archaeology and the operation, which took place over two dive seasons, was criticized for not completing a thorough and careful documentation of the wreck and its contents. After an initial exhibition of a selected number of artifacts from the wreck in Singapore in 2005, a more extensive exhibition was planned for the Smithsonian Museum in Washington, DC. In the run-up to its launch, this was canceled on the back of complaints from archaeologists and anthropologists from the National Academy of Sciences concerning the nature of the salvage operation and the sale of significant parts of the cargo to the Singaporean government for $32 million.[7] Not surprisingly, such issues were rarely raised in Singapore itself, where National Heritage Board and Singapore Tourism Board supported scholarly conferences, publications, and museum exhibitions on the Belitung. Subsequently, the city's Asian Civilisations Museum established an ongoing exhibition featuring a model and items salvaged from the wreck. As an artifact of regional exchange and trade networks, the Belitung is critical to the historiography of Singapore and Southeast Asia more broadly. And as Kwa Chong-Guan notes, it unsettles the idea of Singapore as *terra nullius* before the arrival of Stamford Raffles in 1819.[8] Previous accounts of early trade routes have relied heavily on a combination of Arab and Indian literary references, Southeast Asian epigraphic sources, and Chinese dynastic and textual records. The impact of European merchant empires on Southeast Asian trade is also well documented in archives. But as Kwa argues, the excavation of shipwrecks dating from the ninth and tenth century in the South China Sea, Straits of Melaka, and Java Sea profoundly alters how we understand histories of trade in the region and the importance of Singapore therein. His account identifies not just distinctions in ship design but also changes in the nature of products that were traded over the centuries. When seen alongside other archaeological evidence, the Belitung ship and its contents shed light on shifts in the production techniques of metal and ceramics in China and Southeast Asia and the degree to which designs and technologies changed through maritime trade.[9]

Not surprisingly, remnants of the ship itself were also instructive in understanding the dynamics of trade at that time. Samples of the timber were sent to Australia for testing, but with results inconclusive, attention turned to the mode of construction. Particularly noteworthy was the use of coconut-fiber rope to sew together the Belitung's planks, in contrast to the pegs and nails used in the ships of later centuries. Michael Flecker, the chief excavating officer of the Belitung and founder of Maritime Explorations, has cited use of stitching, together with the ship's sharp bow, removable ceiling planks, and a composite iron and wood anchor, as design features to argue that the location of construction was in the western Indian Ocean region, most likely either Oman or Yemen.[10] Debates about the Belitung's design have also formed part of a small but fascinating line of research that has emerged in recent years concerning the histories of dhows in the Indian Ocean and their modes of construction along the coasts of East Africa, eastern Arabia, and South Asia. As Abdul Sheriff argues, dhows were vital technology in the creation of interconnected maritime cultures across the region.[11] The edited volume on maritime technology in the Indian Ocean by Parkin and Barnes, originally published in 2002, is particularly noteworthy for its various studies of techniques of rigging, stitching, and other elements of boat construction in India, Sri Lanka, and Kenya.[12] As a brief aside, similar studies have also been made on the possible design and construction techniques of Zheng He's treasure ships. Computer modeling and beam theory have been employed to suggest that enormous 450-foot vessels could have withstood the rigors of open sea if built with hulls two to three feet thick, an argument that surely bolsters the grandeur of his voyages.[13]

Interest in the Belitung and Arabian dhows has not been limited to academics. In 2008 a fascinating project, the Jewel of Muscat, was launched. A collaboration between the governments of Oman and Singapore, the project set about reconstructing a Belitung replica. Built on a beach in Oman with timber felled from Ghana, the ship was officially launched during a naming ceremony held in early 2010 led by Goh Chok Tong, former Singaporean prime minister, and Abdul Aziz Rawas, cultural adviser to Qaboos bin Said al Said, sultan of Oman. The aim of the project was to "retrace" the outward voyages across the Indian Ocean that were made by dhows such as the Belitung. A voyage lasting just under five months involving visits to Kochi in southern India, Galle in Sri Lanka, and Penang and Port Klang in Malaysia ended in great fanfare in Singapore. Upon its arrival, Singapore's president, S. R. Nathan, a close follower of the project, officially accepted the Jewel of Muscat as a gift from the sultan to the people of Singapore.[14] Just over a year later, the dhow was placed on permanent display in the Maritime Experiential Museum in Singapore. Both countries sought to maximize the diplomatic benefits of the project, with television documentaries, public events, and publications produced to commemorate the reenacted voyage. Indeed, the subtext of the project was the building of bilateral goodwill between two outward-facing economies located on either side of the Indian Ocean. Celebrations to mark the project's five-year anniversary in 2015 expanded these diplomatic geographies. In a speech given at the Diplomatic Club in Muscat, Oman's secretary-general of the Ministry of Foreign Affairs, Sayyid Badr bin Hamad bin Hamood Albusaidi, sought to frame the project within a history of maritime trade and exchange that had distinct overtures to the new multilateral possibilities of Belt and Road:

> The Indian Ocean played a pivotal role in the development and well-being of the people who live along its coasts. Moreover, the shifts that have taken place over recent decades confirmed the cultural and economic importance of the Indian Ocean as a center of commercial activity and creativity. The Indian Ocean could genuinely claim to have been the birthplace of globalisation from ancient times. Indeed it was across the Indian Ocean that Omani merchants and scholars traveled between multiple cultures, interacting and engaging with one another freely and easily. It was across the Indian Ocean that Oman's relations with many cultures were developed and continue developing today, as they will flourish into the future.
>
> Therefore, it was natural that the Indian Ocean has become one of the most important spheres of activity for Omani foreign policy, in the context of modern-day globalisation. We have consistently sought to continue the spirit of free exchange and partnerships that prevailed during the first era of globalisation. These activities and initiatives continue today, to provide a climate that helps increase cultural, commercial and scientific cooperation, as well as enriching the pillars of coexistence, peace, harmony and mutual respect, so that fair winds may blow into the sails of everyone with productive ideas, for the sake of mutual benefit and common interests.
>
> Singapore has of course been a particularly excellent partner in this regard—not just on the Jewel of Muscat project itself, but in the larger collective project of maximizing the benefit of the Indian Ocean and the Silk Route, in partnership with China and the Indian Ocean Rim countries, and with the excitement of new opportunities and renewed relations. Therefore, I want to take this opportunity to express our deep appreciation for this wonderful and positive collaboration.[15]

Three weeks later Oman turned to India, seizing the anniversary of six decades of diplomatic relations as an opportunity to use histories of maritime sailing to align with the strategic shifts then occurring in India's foreign policy. In 2014, India launched a counter to China's Maritime Silk Road, Project Mausam, an initiative that uses Indian Ocean monsoon histories as a frame for building regional trade and security alliances across its region. Through the involvement of the Indian Ministry of Culture, Project Mausam formed part of the wider political context for the celebrations with Oman. To mark sixty years of bilateral relations, two historic tall ships, the *Shabab Oman-II* and the Indian navy training vessel *Tarangini*, sailed alongside each other for a voyage that would "retrace the historic spice trade route by the ancient dhows between India and the Persian Gulf."[16]

Given the paucity of information regarding histories of sea-based trade in Southeast Asia and across the Indian Ocean more broadly, the discovery of each new shipwreck helps address the imbalance in how the two Silk Roads are seen relative to one another.[17] In his detailed account of maritime archaeology in Southeast Asia, John Miksic highlights the major wrecks that have been salvaged in recent decades and the contribution they make to our understanding of the importance of the region in global trade stretching over two millennia. The salvaging of the Java Sea Wreck back in 1996, for example, provided a glimpse into the scale of commercial shipping of the thirteenth century. The ship sank with around 209 tons of rectangular iron bars destined for Southeast Asian blacksmiths, along with bronze figurines, mirrors, ivory, copper, and tin ingots, as well as more than 110,000 ceramic items.[18] In the following years, two ships dated to the tenth century, the *Intan* and *Cirebon*, were discovered off the eastern coast of Sumatra and northwestern coast of Java, respectively. These would have been sailing across Southeast Asia's waters at a time when the Srivijaya kingdom, discussed in detail in the following chapter, was at its height. Their cargoes of glass, metalware, ceramics, and coins revealed the long-distance trade connections of that period, with items believed to have been manufactured in Persia, Thailand, China, and Java all found on board. As Miksic suggests, they provide critical testimonies to the "integral nature of the China–Indonesia–Indian Ocean trade" at that time.[19] It has been estimated that the *Cirebon* alone was carrying around half a million items. These and other finds indicate the importance the Straits of Melaka held in a long-distance maritime economy stretching back more than a millennium.[20]

Our knowledge about the complex maritime networks of trade that stretched over centuries remains far from complete. Each of the shipwrecks recovered has added to the picture in significant ways. It is clear that maritime trade created a multitude of long-distance connections, but detailed accounts of the scale and geographical scope of contact at particular moments in history are still to be written. One commodity in particular, ceramics, has proved particularly valuable to the recovery of these lost histories. As any archaeologist will testify, shattered into pieces or preserved intact, ceramic artifacts provide a wealth of information about the technologies, social structures, cultural practices, and degree of cross-cultural contact of a group or society. More than any building, they reveal the long-distance connections afforded by the Silk Road of the seas, to use Miksic's term, and the connections that Southeast Asia made with the Roman Empire, the Islamic and pre-

Islamic Middle East, and China. Indeed, Miksic's account of the multitude of trade connections that defined Southeast Asia over a millennium relies primarily on a reading of the travel of ceramics across open waters and the ways this evolved over time as ideas, technologies, and people moved across great distances. In other words, ceramics reveal histories of movement and exchange both through their found locations and in their design and composition.

In Southeast Asia burial sites in West Java dating from the early centuries of the Common Era include pottery "sherds of Romano-Indian rouletted ware," which are likely to have been imported from southern India.[21] The chapter that follows situates such finds in relation to the trade connections established across the Indian Ocean and up into the Mediterranean at that time. Archaeologists have also found small amounts of glazed earthenware of Middle Eastern origin in Thailand.[22] But one of the key insights from the shipwrecks found in Southeast Asian waters is the scale and immense diversity of ceramic items exported from China by ship over several centuries. A deep history of Chinese expertise in ceramic production means that items have long been transported by land and sea to reach admiring audiences in Europe and across Asia. China's unrivaled ability to produce porcelain, a ceramic that requires the high firing temperatures of 2,100–2,600 degrees Fahrenheit, was admired throughout the world and would influence other ceramic traditions far and wide. For hundreds of years, one town in particular, Jingdezhen, dominated the global porcelain industry. Robert Finlay traces the cultural and trade connections that porcelain enabled and the distinct ways in which long-distance exports influenced its production in Jingdezhen, as artisans innovated to meet the aesthetic tastes of Persian and European buyers. The opening of western China during the early Tang dynasty (619–907 CE) led to a flourishing of creativity and the exporting of significant quantities overland, as Finlay explains:

> Oasis communities on the Silk Road played the role of middlemen, conveying their versions of Indian and Persian pictorial methods to China, such as rhythmic patterns, rotating arabesques, stylized flowers, geometric shapes, molded relief, interlaced designs, and exuberant colors. A number of plant patterns, including the acanthus, palmette, and peony-like blossoms, entered the mainstream of Chinese art after monks and artisans copied them from hundreds of Buddhist cave temples and monumental tombs at Dunhuang, near the eastern terminus of the Silk Road. The quintessential Buddhist motif of the lotus, journeying from South Asia to the

Middle Kingdom by way of Persia and the caravan track, embarked on its triumphal progress in Chinese art and architecture. Foreign influence led to one of the most enduring Chinese decorative devices on porcelain plates and bowls, a swirling band of flowers surrounding a central medallion, such as a sketch of a carp, duck, or blossom—a format whose remote descendant is the conventional pattern still frequently painted on modern dinner plates.[23]

Tang-era ceramic camels were among the ornamental items carried across Central Asia, with models of the two-humped Bactrian camels of the period recovered from tombs and caves now commonly celebrated as the quintessential Silk Road motif. As we saw with the wreck of the Belitung, during the Tang dynasty great quantities of porcelain were also exported by ship to Persia and the Arab world. Exports by sea continued to grow during the Song dynasty (960–1279 CE). The bulky, heavy nature of ceramics meant they were more suited to ships than to camel caravans, such that tens of thousands of items were placed at the bottom of cargoes to provide ballast for the long voyages. Song-era ships traveling south from southern China benefited from a body of knowledge concerning routes, monsoon winds, and stopping-off points that had been built up by merchants traveling outward from the Roman Empire and Persia over a thousand years. Discoveries such as the *Intan* wreck have revealed the scale of ceramics production in China in the tenth century. Items from Jingdezhen were stacked in the hold alongside porcelain bowls, jars and ewers made in the provinces of Guangdong, Fujian, and Zhejiang.[24] Shifts in the style and philosophies of ceramic production over extended periods also tells us much about China's engagement with its wider region.

A desire to reassert internal stability during the Southern Song period (1127–279 CE) led to a reversal of a willingness during the Tang era to absorb the external cultural and religious influences that accompanied long distance trade. The elaborate and colorful designs Finlay describes were replaced by a more austere, restrained aesthetic. Lines were simpler, glazes and colors more muted, and although these cool-toned, monochromatic designs of the Song have come to be regarded as some of the finest ceramics ever produced, they also reveal a Chinese culture defined by prolonged battles with more powerful neighbors to the north and a resistance to foreign influence (fig. 5.2). The Java Sea Wreck provides evidence of a great expansion in commercial trade from China in the second half of the thirteenth century, with the ship most likely having sailed around the beginning of the Yuan

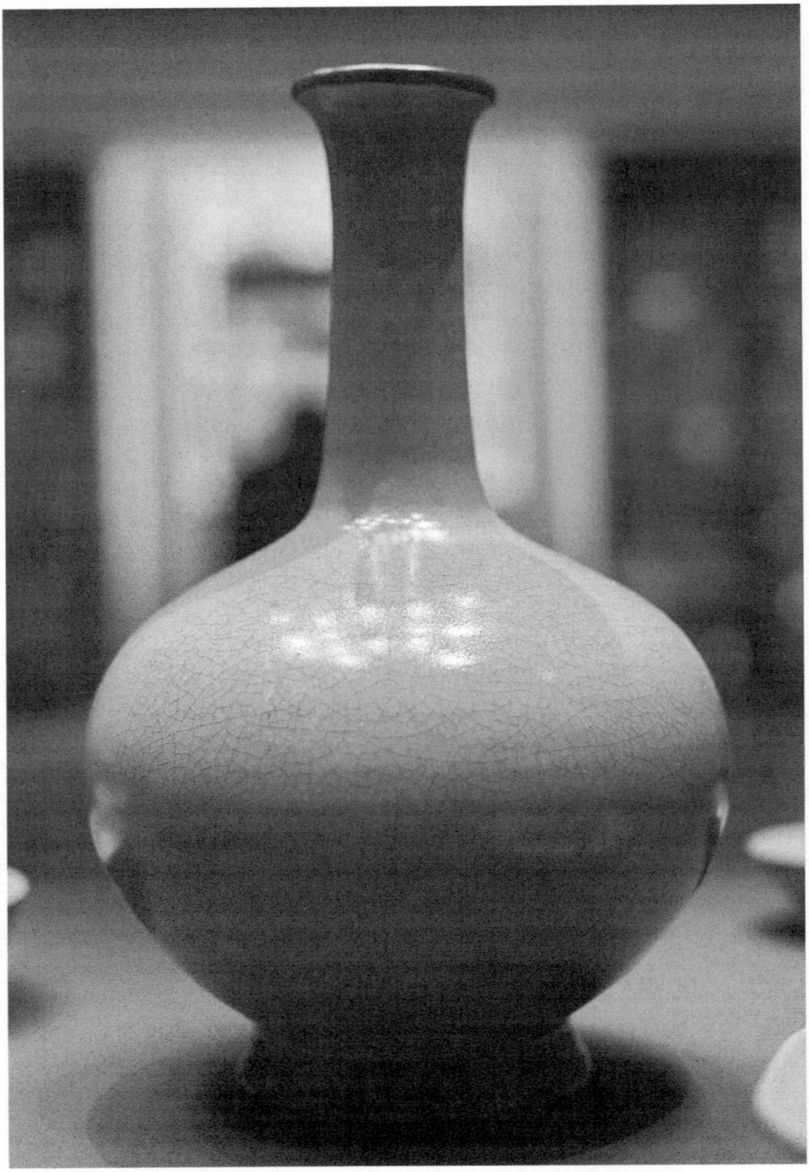

5.2 Bottle with copper-rim mount, Ru ware, Northern Song dynasty, Sir Percival David Collection, British Museum, London. Photo by author.

dynasty (1271–1368).[25] More than one hundred thousand Chinese porcelain items were recovered from the wreck in 1996, as well as a smaller number of items believed to have been produced in Thailand. Of the Chinese items, it is likely they once again traveled from different parts of southern China. The cargo was also noteworthy for its diversity, with utilitarian and high-quality ceramics demonstrating the cross-fertilization of ideas between China and Southeast Asia.[26]

The arrival of the Ming era in 1368 marked a return to a more elaborate ceramic culture in China. The period also saw Jingdezhen consolidate its dominance in ceramics production, with numerous items made for a court that was seeking to regain control over the arts and create more unified forms of cultural expression. During the Ming period, ceramics were exported on an unprecedented scale, with the now-famed blue-and-white designs gaining notoriety among connoisseurs and collectors in Europe. As one of the tangible outcomes of centuries of exchange between Persia and China, blue-and-white ceramics reflected a fusion of Islamist and Chinese decorative aesthetics, a highly distinctive style that relied on the import of cobalt ore from Persia to create the blue pigment (fig. 5.3). The discovery of three pieces of "Tang blue-and-white" on the Belitung was an especially important find, as they are considered among the oldest such ceramics ever found. As Finlay

5.3 Brush and pen holder, Jingdezhen, Ming dynasty, Sir Percival David Collection, British Museum, London. Photo by author.

suggests, the products that were being produced for export to India, Europe, Egypt, Iraq, and Persia during the Ming period centuries later were thus the "product of Eurasian cross-cultural contact, a representative climax to centuries of long-distance interaction and mingled traditions."[27] His account details the various ways in which shifting styles of decorated ceramics both reflected and contributed to this cross-cultural exchange. The Islamic world of southwestern Asia valued an aesthetic of symmetry, rectilinear pattern, and a structuring of space based on mathematics, as evidenced in the architecture and rugs produced in the region.[28] In stark contrast, the Chinese favored the more figurative forms of plants and animals, and those items associated with the religious practices of Buddhism and Taoism. Interestingly, the need to tailor to different export markets also influenced domestic tastes. Equally, potters in the Islamic world took inspiration from the design traditions of Chinese ceramic art, often by transforming figurative features such as birds, clouds or flowers into more abstract, even geometric, forms: "Chinese craftsmen portrayed the dragon as a dynamic creature, an emblem of primal energy, whirling amid clouds or chasing a flaming pearl (a Buddhist symbol of perfection) across the sky; but Mamluk potters and tile makers, oblivious to the Chinese tradition, employed the motif of the dragon as a recurring decorative feature, statically flanking a series of indistinguishable phoenixes."[29]

The chapter that follows details the changing nature of trade in the Indian Ocean. It cites analyses that view the region as a series of world systems whose networks of trade and centers of power waxed and waned over time. As maritime connections carried Islam eastward to South and Southeast Asia, ceramics represented an important commodity. Chinese exporters took advantage of the maritime routes established in Southeast Asia to capitalize on markets farther west, across the Indian Ocean. As we will see, Melaka emerges as an important intermediary of trade in Southeast Asia, with cargoes of porcelain and other items shipped from China offloaded on to ships ready to return to various home ports across the Indian Ocean. Although Chinese ceramics had long been exported across East Asia to Japan and Korea, their biggest markets lay in West Asia, with Ottoman Turkey and Safavid Iran both developing an appreciation for high-quality blue-and-white. As part of her account of the global consumption of Ming porcelain, Stacey Pierson describes the absorption of ceramics into the Safavid Iran (1501–1722), arguing that the appropriation and new meanings given to Chinese designs needs to be read as a form of cultural hybridity.[30] Until the mid-fourteenth century, Iran influenced the styles and technologies of other potters in the Middle East. But the arrival of blue-and-white ceramics from China starting in the early fifteenth century would profoundly alter production across the Islamic world for centuries to come. Some items traveled overland, but Hormuz and Gombroon (Bandar Abbas) became key entry points, where cargoes were unloaded throughout the sixteenth and seventeenth centuries.[31] As William Honey explains, vast quantities of blue-and-white and monochrome blue-, green-, and brown-glazed items traveled from China to Persia at this time.[32] More recently, Lisa Golombek and Patty Proctor have elaborated further on the influence these items held on Iranian pottery. Their detailed analysis of the production of different workshops across the country reveals shifts in design motifs and shapes in response to changing aesthetics in China.[33] In the mid-sixteenth century, for example, flowers and rims comprising scrolls and spirals were extensively imitated in workshops in Tabriz, in northern Iran. Farther south, in Kirman, seventeenth-century potters copied Chinese calligraphy and landscape scenes featuring birds, deer, and insects. The fine craftsmen of the workshop also creatively adapted China's iconic motifs, as the description by Lisa Golombek of a dish from that period indicates: "The dragon's body is transformed into a rigid circle to conform to the shape of the tondo, but the naturalistic treatment of the reptile's skin animates the dragon, recalling the dramatic renderings of dragons in foliage in the marginal illuminations of Persian books."[34] Farther east, Ming porcelain was also consumed in significant quantities in Mughal India. Pierson describes the presence of Chinese ceramics in Indian miniature-painting banquet scenes, including one of an Iranian ambassador being received by Akbar (1542–1605). The accounts of Ibn Battuta and other literary references also testify to the widespread appreciation of Yuan-era and early Ming-era pieces as antiques. Precious pieces were offered as diplomatic gifts, such that items produced in the fifteenth century would become the cherished artifacts of the seventeenth century.

Europe's "discovery" of the New World and Vasco da Gama's establishment of a sea route to India via the Cape of Good Hope in 1498 greatly expanded the markets for Ming porcelain. The high suitability of porcelain for sea transport meant that it could be successfully carried great distances. As noted earlier, because of its bulk and weight, it served as an ideal form of ballast on rough seas. But unlike textiles and items crafted from wood, ceramics were resilient to the corrosive effects of seawater, and in contrast to spices, the lack of odor meant they could be safely carried alongside tea. Records documenting the tableware of royalty point to the value placed on Chinese porcelain in fifteenth- and sixteenth-century Europe. But it was the establishment of a Portuguese base in Macau that would mark the beginning of a new era of export westward from China to the Middle East and to markets in Europe and the Americas beyond.[35]

The battle for naval supremacy and the spoils of looting in Asia would have long-term ramifications for everyday cultural life in Europe. In the early years of the seventeenth century, hostilities between the Portuguese and the Dutch involved attempts by the latter to capture the cargoes of Portuguese vessels en route to and from Goa and Macau. In 1603 the *Santa Catarina,* an enormous ship for its day, surrendered to Jacob van Heemskerck, the Dutch commander of a fleet of eight vessels. After a daylong battle, the Dutch took ownership of a cargo of thousands of bales of silk and an estimated 132,000 pounds of porcelain.[36] With this and subsequent bounties sold on the Dutch market, Chinese goods began to make inroads into Dutch life. The manner of acquisition also lent itself to the title given to the imported porcelain. The Dutch word for the Portuguese carrack galleons, *Kraak,* led to the widespread adoption of the term *kraakporselein.*[37] The term has come to designate a particular design style whereby panels are laid out in various arrangements. More commonly adopted for flatter items such as plates and shallow bowls, panels depicting decorative motifs—plants, flowers, animals—were interspersed with narrower panels containing geometric patterns. In this *Kraak* porcelain, then, we once again see the fusion of principles of design from the Islamic world and China itself. Interestingly, the Dutch were also importing pieces that featured family coats of arms and design modifications to suit Portuguese dining and drinking culture. The Dutch East India Company (or VOC, for Vereenigde Oostindische Compagnie) flooded the market, and in less than two decades after the capture of the *Santa Catarina,* the Dutch market was reaching saturation point. Chinese merchants sailed cargoes of *Kraak* porcelain to the Dutch ports of Bantam, Batavia, and Patani, with Dutch ships carrying vast quantities back home to Europe. The delicate nature of *Kraak* porcelain, its thin light walls, also meant it gained popularity across Southeast Asia and Iran. VOC ships regularly stopped off at Gombroon, on the southern Iranian coast, but as with the Portuguese and English that followed, such ships brought silks and spices from India, ceramics from China and Japan, and weapons from Europe.

To reduce the risk posed by pirates and smugglers and to ensure a regular supply from the factories of Jingdezhen, the VOC increased its presence on the island of Formosa, building a stone fortress in the

1620s to secure its position in the region.[38] After several years of conflict with Chinese authorities regarding trade, relations picked up after 1633, with Formosa serving as a vital intermediary between Japanese and Chinese merchants given that the former were denied permits to trade on the Chinese coast. As Moster and Van Campen explain, proximity to Jingdezhen gave the Dutch more control over the products manufactured for export. Although the design motifs used in previous decades remained common, the types of items produced underwent change, with beer mugs, vases, and condiment pots making their way into the holds of ships bound for Amsterdam. Over the course of several decades, the VOC imported millions of items from China. These in turn spawned a domestic, more affordable ceramics industry with the characteristic blue-and-white decorative style remaining the signature of Delftware for more than two hundred years.

In this brief overview, the various examples demonstrate how the aesthetic and cultural expressions that ceramics are imbued with testify to mobility, and thus to absorption and integration, hybridity, and ongoing exchange and modification.[39] Pigments and glazes, along with firing techniques and mineral compositions, all tell the story of complex itinerancies—they are pieces in the jigsaw puzzle of how objects followed the routes of trade and found temporary residency in market towns and the warehouses of maritime ports. Even in their very location of recovery, ceramic items such as those found on the Belitung, the Java Sea Wreck, or the *Intan* give us insights into the cores and peripheries of the Silk Roads. Resilient to saline and safely protected from centuries of warfare, invasions, revolutions, and the everyday accidents of domestic life, shipwreck ceramics are among the fragile but surprisingly durable items handed down to us. What we begin to see through them is how the value, even authenticity, of ceramics can lie more in routes than in roots. They reveal how technologies, concepts, and designs migrated and evolved in response to aesthetic tastes that were held locally or at great distance, and they offer valuable insights into the intersections between maritime and land-based routes. But as Finlay notes, ceramics also operate at a certain level of "cultural abstraction and metaphor, closer to sculpture and painting than to salt and sugar."[40] And it is these different attributes that are now being mobilized by the cultural and political agendas of Belt and Road. As we are about to see, they are among the artifacts of previous centuries now reassembled in particular ways, performing a distinct role in the heritage diplomacy of Belt and Road. It is a situation that means they are also taking on twenty-first-century itineraries of travel, as they move between the growing number of museums across the region that are dedicating exhibition space to the history of the Silk Roads.

Exhibiting Itinerancy

Over the past twenty years, cities across Asia have invested significantly in building and renovating public museums. They have become valuable cultural infrastructure assets for cities or regions looking to gain competitive advantage in increasingly competitive tourism and international labor markets.[41] Much attention, some of it critical, has been given to the so-called Bilbao effect and the capacity for museums, cultural quarters, and "starchitecture" to serve as catalysts for economic regeneration.[42] Less well developed, however, is our understanding of the role of museums in international relations and cultural diplomacy; studies by Sylvester and by Lord and Blankenberg represent notable attempts to conceptualize this issue.[43] The latter consider museums as agents of soft power, highlighting how cities and governments use them to signify a set of values or levels of social and cultural attainment: democracy, peace, innovation, human rights.[44] As in the examples of Xi'an, Dunhuang, Lamu, Galle, Melaka, and Singapore, museums are playing a distinct role in the public diplomacy agendas of Belt and Road, reaching both domestic and international audiences. Indeed, since the launch of Belt and Road there has been a marked increase in the number of exhibitions and the volume of museum space dedicated to Silk Road histories.

To explore this issue further, I return to the question of maritime archaeology and the South China Sea. As highlighted in previous chapters, China's maritime archaeology institutes—most notably the National Center for Conservation of Underwater Cultural Heritage—have deployed ships and submersibles equipped for deepwater shipwreck hunting in an attempt to find material evidence in support of claims for territorial jurisdiction over the East and South China seas based on historical right.[45] In contrast to many commentaries by Western experts that point to twentieth-century laws and treaties as the historical context of the dispute, wrecks containing porcelain and other artifacts dating as far back as the ninth century help substantiate Beijing's position. But with the discourse of the Maritime Silk Road shifting the argument from claims of historical presence to the more diplomatically expedient language of regionwide trade, encounter, and exchange, the National Center for Conservation of Underwater Cultural Heritage has reframed its work to align more explicitly with the strategic agendas of Belt and Road. As Jiang Bo, director of the center's Institute of Underwater Archaeology, stated in 2015, "The country is gradually turning towards the ocean for a development strategy; the Belt and Road initiatives surely give an impetus to our studies."[46] For Jiang Bo, then, excavating evidence of maritime travel and contact offers clear lessons for those working on developing China's strategic influence in the region today: "For all the research I've been doing on the Maritime Silk Road, I'm most amazed at how much we're shaped by the communications (with other cultures). . . . We found that flourishing ancient civilizations never keep to themselves. And China has been most influential when it is opening up and communicating with other cultures."[47]

Of course, one of the key reasons that states fund such forms of archaeology is to provide content for museums. Indeed, since the early 2000s China has been eager to acquire ceramics from the South China Sea for public display. In 2004 this led Christie's Hong Kong to hold an auction of seventeen thousand Ming-era Chinese ceramics in Melbourne. Aware of the intense interest in the collection from buyers associated with the state in China, their decision to relocate to Australia was an attempt to depoliticize the sale of a haul recovered close to the Vietnamese coast.[48] Some years later, the Guangdong Maritime Silk Road Museum opened its doors to the public to show its significant collection of recovered ceramics. Not surprisingly, the launch of Belt and Road provided the impetus for further museums and temporary exhibitions related to maritime heritage. In late 2014 nine cities in China gathered together their collections for an exhibition titled *Over the Sea*, which moved on to several cities after its initial launch in Quanzhou. Porcelain, temple stones, and silks were among the items put on display at a time when the city was in the advanced stages of planning a UNESCO World Heritage Site nomination for the Maritime Silk Road.[49] The exhibition also represented a significant step toward a longer-term strategy to build a new, permanent museum dedicated to the Maritime Silk Road.[50] With Jinjiang designating itself as the city "where the Maritime Silk Road began" and Nanjing renovating its Zheng He treasure-ship boatyards and museum facilities, we thus see the emergence of new forms of exhibitionary competition. Indeed, Ningbo took the somewhat-unconventional strategy of deploying artifacts in glass cabinets in subway stations and decorating train interiors with Maritime Silk Road–themed cartoons.[51] Such exhibitions form part of civic education programs, with local residents—most explicitly parties of schoolchildren—presented with a variety of object histories that connect their home city with other locations in the region.

In 2016 the National Silk Museum in Hangzhou closed for a major renovation, as did the Maritime Experiential Museum in Singapore a year later, with promises that when they reopened, guests would enjoy "an all-new and immersive gallery . . . as they journey through the Maritime Silk Road." The same year the Hong Kong Maritime Museum also inaugurated its new research center dedicated to the Maritime Silk Road. Tourism provides the most tangible mechanism by which these museums reach international audiences. But as the Jewel of Muscat and Xi Jinping's visits to museums in Uzbekistan and Georgia illustrate, museums can also be sites of diplomatic hospitality, whereby the receiving of foreign political elites is the subject of media coverage in both countries.[52] In 2016 China and Egypt embarked on a series of "civilized conversations" as part of the so-called Sino-Egyptian Culture Year. Exhibitions bearing the title *Silk on the Silk Road* were held in both Luxor and Cairo, with display collections supplied by the China National Silk Museum and Zhejiang Cultural Centre, Hangzhou.[53] From there, the exhibition traveled to Doha, where Princess Sheikha Al-Mayassa, sister of the ruling emir and also chairperson of Qatar Museums, declared it open. Once again, *Silk on the Silk Road* explicitly invoked, but with few specificities, the historical connections of the overland and maritime Silk Roads as the basis for building dialogue and friendship in the twenty-first century (fig. 5.4).[54]

Underpinning these various examples is a distinct political logic by which artifacts, invariably sourced from several locations, come to be assembled. In this regard we see them take on new itineraries as they move between different social settings and acquire new symbolic values. Perhaps most interesting, though, are the traveling exhibitions, in that their objects come to act as diplomatic agents as they travel. The silk items displayed in Cairo, Luxor, and Doha follow carefully chosen itineraries of heritage diplomacy as they travel to locations of strategic value to Beijing, namely its Belt and Road partner countries. The long-standing practice of galleries and museums sending collections on traveling exhibitions has often revolved around first-tier cities with the largest audiences, namely London, Los Angeles, Sydney, and Paris. And while many roaming exhibitions on the overland Silk Road have graced the halls of the West's most prestigious cultural institutions, Belt and Road has led to some fascinating shifts in these geographies

5.4 *Silk on the Silk Road* traveling exhibition, Doha. Photo by author.

of cultural diplomacy, as new itineraries for silk, ceramics, and stone appear. But if we project this trend into the future, shifts in the geographies of collecting and display come into view, ones that are likely to cause deep concerns over the coming years.

Trafficking Antiquities

One of the most shocking developments of the conflict in Iraq and Syria since 2011 is the looting of archaeological sites and museums. In addition to the attempted cultural erasure by destruction, Islamic State extensively trafficked antiquities to build a stream of revenue for the organization. As IS began to occupy significant territories in Syria in late 2013, it capitalized on an existing looting industry by imposing taxes on the sale of artifacts.[55] The Syrian scholar Amr Al Azm has traced Islamic State's issuing of digging permits and penalties for non-payments, as well as its ability to take greater control of the process through the second half of 2014. To better organize the export of items out of the country, the organization established the Manbij Archaeological Administration in Manbij, a city close to the Turkish border.[56] This represented a more systemic, industrial approach to looting, incorporating numerous sites and museums across both Syria and Iraq, and the use of dealers and intermediaries to extract profit as artifacts left IS-controlled territories.[57]

In the attempt to combat this illicit trafficking of cultural artifacts, debate has long surrounded the importance of tackling both the supply and demand sides of the problem.[58] In 2015 the International Council of Museums published the multiauthored *Countering Illicit Traffic in Cultural Goods*, placing the spotlight on international networks of dealers, collectors, and museums—as well as themes like provenance and documentation—to lay out the complex, often opaque nature of the conflict-antiquities market today.[59] One of the recurring themes of such analyses has been a particular transnational geography of supply and demand. Over recent decades a small number of countries—namely those that are home to former great cultures or civilizations and suffered major social upheaval in the modern era—have furnished the international market with the majority of illicitly trafficked artifacts. Examples include Colombia, Cambodia, Libya, and of course Syria and Iraq. Overwhelmingly, items from these countries have flowed westward to the wealthiest nations of Western Europe or the United States, with London and New York becoming principal hubs in a network of networks involving auction houses, collectors, and dealers. It is a transnational geography that involves particular corridors of smuggling, whereby items cross multiple land and sea borders in Central and South America, or up through North Africa and southern Europe. In raising concern about the "war chest" that the Islamic State was building from looting, various investigative reports published in the Western media highlighted the high prices dealers and collectors in the United Kingdom, United States, France, and elsewhere were prepared to pay for smuggled artifacts.[60] There remains a dearth of precise data concerning the total revenue generated by these sales, but the looting of individual sites has been estimated to have generated Islamic State approximately $40 million, leading various counterterrorism experts to argue that illicit trafficking of antiquities was a significant obstacle to diminishing the organization's capacities.[61]

Yates and colleagues argue that important shifts are taking place in these international geographies of trafficked antiquities. East Asia is joining Europe and North America as a destination market. I would argue that the "revival" of the Silk Roads is likely to accelerate this trend over the longer term.[62] Former Silk Road cities in the Middle East—Palmyra, Aleppo, Bosra, and Damascus in Syria, and Hatra, Erbil, and Mosul in Iraq—could become the source of new forms of trafficking, as museums and private collectors in East Asia and elsewhere in the region seek to build collections. Since the mid-2000s China has made staggering investments in the construction of new museums. With hundreds opening every year as part of a five-year strategy to create a national infrastructure of cultural institutions, a target of 3,500 new museums by 2015 was exceeded three years previous.[63] Such ambition raises questions over the collections that will be required to fill vacant exhibition space.[64] As thousands of curators across the country face this challenge, it is highly likely that a certain proportion of them will look to the Silk Roads for content. Museums have become key components of strategies pursued by municipal and provincial authorities to culturally locate their cities within the new national imaginary of a culturally vibrant and strong China. As they compete for central government funding, associations with Belt and Road, as signaled by connections to the "ancient" Silk Road, hold major strategic value. In other words, museum acquisitions policies will form part of larger strategies adopted by cities seeking to showcase their historical connections to other Silk Road locations in China and beyond. It remains to be seen how many of these institutions acquire illicitly trafficked items.

A trend likely to have an even greater impact, however, is the acquisition of artifacts by private collectors. Michael St. Clair traces a tradition of private collecting in China stretching back more than a thousand years. Antiquarianism, he suggests, has "had a close connection with loyalty to the state."[65] Over the past 150 years or so, however, China has experienced an extraordinary loss of art and artifacts. Wars, invasions, and revolutions have caused immeasurable damage, and it is estimated that more than ten million cultural objects have left the country for Europe, Japan, Taiwan, and most notably the United States.[66] The sharp increase in the value of Chinese art during the 1990s extended to antiques, with a growing cultural elite pushing up demand and prices. More recently, Xi Jinping has tied antiques into a national agenda of revival. As Gallagher notes, in 2015 Xi urged local administrators and cultural bureaucrats to instill a sense of national pride and patrimony over the past.[67] Crucially, China has become one of the world's largest markets for luxury goods and is home to the fastest-growing population of US-dollar millionaires.[68] Manuscripts, artworks, and other archaeological artifacts constitute a particular form of exclusivity, allowing owners to display and enact the forms of "cultural capital" made familiar by Pierre Bourdieu.[69] As Denis Byrne has noted in the context of Southeast Asia, the private collecting and connoisseurship of antiquities in the region over the course of the twentieth century emerged in tandem with the enhanced prestige given to objects via fields like archaeology and art history. As laws are ratified and states introduce formal collecting practices for museums, the level of private collecting goes up. In Southeast Asia, the boundaries between state and private collector have been further blurred, with members of the political elite, bureaucracy, and business class among those wishing to display their refined taste and aesthetic appreciation through collections of antiquities.[70] There are good reasons to assume a similar dynamic will continue in China around the Silk Road, as a multitude of collectors, both public and private, mutually reinforce and escalate the symbolic, political, and financial value of its material culture. In a reversal of the destruction of private collections during the Cultural Revolution, a vast market of private collectors is likely to emerge using an array of formal and informal channels, ranging from Christies and Sotheby's in Shanghai and Beijing, respectively, to black-market dealers.

The structures of Belt and Road will compound the issue. In the development of its overland economic corridors, we are effectively seeing an infrastructure of smuggling being built. The roads, rail lines, and opening up of borders through Pakistan, Turkmenistan, Uzbekistan, Tajikistan, Kyrgyzstan, and Kazakhstan means that multiple corridors will form along which items can be illicitly trafficked from the Middle East into China. As cross-border trade is actively encouraged by governments across the region, myriad challenges arise in the effort to track and recover cultural items. It is likely that many of the international mechanisms introduced in recent times, such as the Red Lists published by the International Council of Museums, as well as the 1970 UNESCO Convention on the Means of Prohibiting and Preventing the Illicit Import, Export and Transfer of Ownership of Cultural Property and the 1995 UNIDROIT Convention on Stolen or Illegally Exported Cultural Objects will be of limited value in what is a highly porous, and in places "lawless," part of the world.[71] In Central Asia, only Azerbaijan, Kazakhstan, Pakistan, Russia, Tajikistan, and Uzbekistan have ratified the 1970 convention. UNIDROIT was designed to address gaps in the 1970 convention by establishing conditions for restitution and obligating buyers to check the legitimacy of their purchase. Neither Armenia, Kazakhstan, Kyrgyzstan, Tajikistan, Turkmenistan, nor Uzbekistan has signed or ratified the convention. If the illegal trafficking of cultural items during times of conflict in the Middle East continues, this combination of new corridors of smuggling and a fast-growing and potentially vast market for cultural artifacts in China and the region will pose a series of whole new challenges for the international preservation sector.

SIX

Historical Openings

Chapter 2 traced the rise and international circulation of the Silk Road concept. This begs an important question concerning whether a global awareness of a metaphor for exchange, trade, and dialogue produces new understandings of Eurasian history. I would argue Belt and Road constitutes a political economy that, over the longer term, will enhance the visibility of, and give form to, historical events and topics that have not received the attention they deserve. Precisely how this will unfold remains a complicated question. This chapter sketches out some possible road maps by exploring questions of world history and the narratives of historical and cultural significance, which have formed around nations and cities across the region through discourses of heritage. The idea of the Silk Roads provides an all-encompassing frame for making visible an array of events and historical processes that have important implications for how we think about world history. As scholars such as Pearson, Frank, and others have argued, the worlds of Central Asia and the Indian Ocean before the fifteenth century continue to be poorly understood, too often cast as the passive regions of the East and spaces of irrelevancy until the arrival of European capitalism. The idea of reviving both the overland and the maritime Silk Roads thus represents a catalyst for unearthing and shining a light on—and I use those in both a metaphorical and a literal sense—a vast array of artifacts from the past that speak of previously unknown interconnected histories. Extraordinary potential exists for new stories to be told through academia, museums, heritage sites, and the media. Belt and Road holds the potential for recalibrating how the historical significance of Eurasia's cities, coastal and inland, is articulated. But forces and assemblages produced through Belt and Road also close down the discourses of history and heritage in very particular ways.

The substantive analysis offered here focuses on the forms of connection that occurred across the Eurasian land mass and Indian Ocean before the arrival of European naval power in the fifteenth century. Each of these regions has been the subject of increased scholarly attention in recent times, with authors readily admitting that their accounts capture a small part of the story given the shortfalls in primary data sources and the chasms in knowledge that remain about certain topics and events. In citing such works, my aim is to be neither comprehensive nor prescriptive. Rather, I have selected themes deemed most relevant to the possible paradigm shifts Belt and Road will bring about. Chapter 2 highlighted how the concept of the Silk Roads took on different inflections in different countries over the course of the twentieth century. For China, as with India and Japan, the Silk Road is a history of intraregional connection as much as it is about the long dialogue between East and West. The Silk Roads of the twenty-first century put China at the center of that story. But as we have seen, as questions of Silk Road pasts are taken up outside China interesting complexities emerge regarding which pasts and points of connection are valorized and celebrated, and which events continue to be ignored. The final section discusses some of the mechanics of this by considering how Belt and Road has created new configurations of academic inquiry and knowledge production.

In terms of the Maritime Silk Road, we have heard much about Zheng He and the port cities and shipwrecks of Southeast Asia and the South China Sea. This chapter deliberately expands the geographic scope to the wider Indian Ocean, and the multivalent flows of religion and culture that occurred therein over a number of centuries, to emphasize the problems of delimiting these maritime histories to a single "road," one that supposedly commenced in the port cities of eastern China and reached its pinnacle with the seven voyages of the Ming era. Clearly, certain forces will shape which stories get told and how the geographies of connectivity are reconstructed via the narratives of heritage. In recognizing this, it is important to not be overly deterministic. Silk Road heritage will be shaped in large part by where the past is unearthed, from which archives, and from which archaeological sites.

But it will also be determined by who undertakes such tasks, to what ends, and who pays to put things on public display. This is the fascinating future of the past now unfolding across Eurasia and beyond.

One of the first indicators of a Silk Road revival has been the renewed interest shown by media companies and academics, two sectors in which a fresh wave of content and knowledge has been produced on the histories of Eurasian trade, cultural and religious exchange, and maritime travel. Belt and Road has given impetus to such material being produced across these two sectors, funded as part of public diplomacy programs and strategic investments made by governments across the region. In Hong Kong, China, Singapore, Australia, and elsewhere, academic funding has been strategically targeted toward research initiatives associated with Belt and Road. As we will see later in the chapter, this is creating fascinating new clusters of academic knowledge production, whereby historians and archaeologists are finding themselves in collaborative settings that would have been unthinkable before Xi Jinping's visits to Kazakhstan and Indonesia in late 2013.

Explorations in History

The idea of the two Silk Roads encourages us to revisit the received wisdoms of world history. In accounts of the emergence of the modern world order, Europe has long been privileged as the preeminent region of change and transformation at every level of society. Structuralist conceptualizations of integrated trade systems, as offered by authors such as Fernand Braudel and Immanuel Wallerstein, have been particularly influential in this regard. Braudel argued that the world economy exhibited foundational structures, and only by looking beyond episodic events and cyclical patterns could we reveal the *longue durée* of history. Inspired by this macroanalysis of geoecological regions, Wallerstein's model of global capitalism began with the major transformations that occurred from 1500 onward and the ascendancy of a maritime Europe.[1] As world-systems theory secured a widespread following from the 1970s, it served as an important explanatory frame of precolonial trade, colonialism, and progress. Although Wallerstein's account of core and periphery zones has been critiqued at length in recent decades, its identification of a European center remains a normative architecture guiding the majority of world-history theses today. At the same time, however, authors working across different geographical contexts and time frames have engaged with his framework, expanding and complicating its conceptions of cores and peripheries by identifying trade entanglements between regions and by tracing non-European contributions to European culture from the fifteenth century onward. Walter Mignolo, for example, mapped the economic and cultural flows that conjoined Europe and South America starting with "the age of discovery."[2]

For those interested in early trade connections across the Indian Ocean and between West and Central Asia, Wallerstein's original thesis suffered from weaknesses in its periodization of capitalism and the importance it placed on precapitalist trade. In his 1998 volume *ReOrient*, Andre Gunder Frank argued that only if we turn our attention to the region of Central Asia do we begin to see the fallacy of Eurocentric historiographies, including those offered by Wallerstein and his followers. By tracing the interconnections between southern Europe and East Asia, and the role of Central Asia therein, Frank called for a recasting of how we geographically locate stories about heights of civilization, centers of technological innovation, and core-periphery trade dynamics. In its historicization, the Indian Ocean shares important parallels with the Middle East and Central Asia. Historiographies of exogenous forces have prevailed, particularly those that traveled from or near Europe. Western scholarship on the Indian Ocean has prioritized understanding the cultural and economic influences the Roman Empire, Islam, and the great age of European discovery brought to the region. This has held significant implications for the ways in which port cities have been understood as historical and heritage landscapes, as we will see shortly. Pearson, Beaujard, and Frank thus suggest that Central Asia and the Indian Ocean have too often been cast as dormant, unchanging, and outside of history. Together they argue that we need to not only better understand the central role played by Eurasia and the Indian Ocean region in world affairs over several millennia but also develop more robust analyses that transcend those transitory moments when external players shaped events.[3] Pinning down the causes of such Eurocentric discourses is no simple task. It would be easy to lay the blame at the enduring fascination and nostalgia Europe has for its maritime empires. But as we saw in chapters 1 and 2, there are various broader factors that have influenced east-west historiographies in the modern era.

Since the 1980s, however, we have seen the steady emergence of a body of scholarship attending to the concept of regionalism in Asia and its merits as historical method. Books by Janet Abu-Lughod and Kirti Chaudhuri opened up new analytical pathways in this regard. Chaudhuri, for example, adopted a mathematical analysis for the question of how cultural and economic connections between cities and regions emerged across the Indian Ocean via a class of mobile merchants. Frank's own work *ReOrient* traced the cyclical trade relations of the global economy for the period 1400–1800, arguing that Asians, and most notably the Chinese, were systematically influential in this period and far from the backward and traditional inferiors to their European counterparts, as they are often cast. Accordingly, he states: "In no way were sixteenth-century Portugal, the seventeenth-century Netherlands, or eighteenth-century Britain 'hegemonic' in world economic terms. Nor in political ones. None of the above! In all these respects, the economies of Asia were far more 'advanced,' and its Chinese Ming/Qing, Indian Mughal, and even Persian Safavid and Turkish Ottoman empires carried much greater political and even military weight than any or all of Europe."[4]

Since then, historians attempting to trace such connections further back have asked whether the Indian Ocean constitutes a single unit of analysis and, if so, where its geographical and cultural boundaries might lie. Michael Pearson explores this question at length, and in an account that draws inspiration from Braudel's analysis of the Mediterranean, he acknowledges that the trend to trace the continental history of Eurasia, which emerged in the 1990s, influenced those interested in theorizing early seaborne relations. One of the great advantages of writing maritime history, he argues, is the opportunity to look beyond states as the normative architecture of analysis and understand how "worlds" or "zones" came about.[5] It is also a disposition toward contact and links rather than territory and frontiers. Pearson highlights the various analytical hoops that need to be jumped through to speak of an interconnected Indian Ocean region. Typically its western frontiers are marked by the East African coast, with evidence of dhows traveling up and down the coastline of the Arabian Sea and Persian Gulf cited as part of a larger history of littoral societies that included port cities in southern India and beyond. The task of demarcating boundaries to the south and east, however, becomes more difficult. Whether the Straits of Melaka represent a boundary point or the crucial link of an extended maritime region that includes the South China Sea remains a point of conjecture. In capturing this geographical and cultural ambiguity, Pearson thus suggests that if "there is a wide, expansive Indian Ocean, around its edges and margins are a host of seas. Among them are the Mozambique Channel, Red Sea, Gulf of Aden, Arabian Sea, Persian Gulf, Gulf of Oman, Bay of Bengal, Andaman Sea, Strait of Melaka, and the Laccadive Sea."[6]

Philippe Beaujard has pursued such a geographically expansive analysis, modifying Wallerstein's world-systems theory to account for precapitalist forms of trade stretching back to the fourth millennium BCE, with Mesopotamia as its core. He argues that as this world system expanded, land and maritime trade routes developed across Anatolia and Central Asia, incorporating Egypt and the Indian Ocean by the second millennium. The emergence of a central Eurasian culture around the time the Achaemenid Empire forms also meant that vast areas of the Northern Steppe were Iranian speaking by the fifth century BCE.[7] In offering a sweeping arc of history, Beaujard cites archaeological evidence to argue a subsequent world system formed from around the seventh century BCE with India acting as its center. Exchanges with China were also established some four hundred years later.[8] By the first century BCE, however, the centers of power had shifted to the Han Chinese and Roman Empires, with both exerting considerable influence on maritime and overland trade across the cities of Eurasia.

Historians disagree over whether these periods and transitions in trade constitute a singular world system or a series of smaller regional ones. For Chew, as the first century of the current era ended a level of integration occurred such that we can speak of the "first Eurasian world economy," incorporating Europe, East Africa, and much of Asia.[9] Beckwith neatly summarizes how the land and sea components of this expansive economy related to each other over the centuries:

> The old maritime trade routes and the continental trade routes thus did not conflict, though the possibility of obtaining goods by more than one route may have exerted some competitive downward pressure on prices. The two existed throughout history, but purely as different subsystems of transportation and distribution within one Eurasian continental trade system, the center of which remained the Silk Road, the Central Eurasian economy. The region where the two routes met and interacted most intensely was Southwest Asia, primarily meaning Iran, Iraq, Egypt, Syria, and Anatolia. To some extent the political power of Persia throughout history is inseparable from its strategic position between East, South, and West by land and by sea. The same is true of Anatolia and Greece, which supported the Eastern Roman Empire, the Byzantine Empire, and the Ottoman Empire.[10]

Archaeological evidence suggests that the Kushan Empire, which emerged around the first century CE across the region known today as Central Asia, evolved as a highly syncretic culture. It integrated nomadic peoples from East and West; absorbed elements of Hellenistic culture, Buddhism, and Zoroastrianism; and maintained contact with the Chinese courts. Ruling Kushan from Purusapura (modern-day Peshawar), Kanishka the Great expanded the empire south into India and briefly took control of important trading cities such as Kashgar and Khotan, both of which lie in present-day Xinjiang. After the Kushan control over the region began to collapse early in the third century CE, no single power exerted its will on the trade routes crisscrossing the mountains and plateaus of the region. Nonetheless, large quantities of silk and other items continued to pass through thanks to autonomous trading networks sustained by religious institutions, merchants' organizations, and local communities. From the Mediterranean to China, traders traveled overland to supply markets with a multitude of goods despite sporadic military conflicts, dynastic changes, and even imperial authorities' attempts to protect and pursue their own trade interests. The markets and products of the Han, Kushan, Parthian, and Roman Empires may have been the source of much trade and cultural exchange, but these continued to thrive long after these empires collapsed.[11]

The prolonged period of political stability created under the Abbasid Caliphate (749–1258) contributed to the peaceful dissemination of Islam across the western Indian Ocean world. As Islam moved out from the Arabian Peninsula to North Africa and through present-day Turkey, Egypt, and Iraq, Arab and Persian seafarers also carried the religion down the East African coast in the eighth century and across to the Indian subcontinent. By the fourteenth century Islam had made significant inroads across Southeast Asia, influencing the customs and everyday life of port cities in Sumatra, Java, and the Malay Peninsula. Distinctions have been made between the absorption of Islam among inland and coastal regions. Its migration across inland regions, in some cases via forceful coercion, influenced the existing social and political order in far-reaching ways. For coastal regions, however, regular contact with long-distance travelers meant that communities were more receptive to Islam's cosmopolitan outlook, in what Ross Dunn has called "a new feeling of participation."[12] As it spread across the Indian Ocean, often remaining a minority faith, the points of connection Islam created continued to proliferate, fueling a sense of cosmopolitanism rooted in, and routed along, affinities between communities living in East Africa, India, and Southeast Asia. While it would be misleading to equate this with the forms of transnational identities that have taken shape in recent times, a number of authors have pointed to the tradition of the hajj as a significant factor helping to create cultural ties between groups linked by long-distance trade and its merchants.

Long before the Islamization of the Indian Ocean, Buddhist pilgrims from China also followed maritime routes to the Indian subcontinent. From the seventh century Buddhism spread across Southeast Asia as the thalassocratic city-state of Srivijaya expanded its influence outward from Sumatra. Drawing on archaeological evidence from a number of locations across the region, John Miksic has identified the extensive trade connections Srivijaya developed with China, South India, and the Arab and Persian worlds.[13] Firmly in control of the Straits of Melaka, Srivijaya benefited from growth in commercial traffic, altering its nature through the development of Melaka as an intermediary point for the transfer of cargo between boats traveling from China and the Middle East. Janet Stargardt draws on material recovered from three shipwrecks in the region to paint a more detailed picture of how this transition in the pattern of trade occurred. She argues that those ships crossing the waters of the Indian Ocean and South China Sea in the ninth century undertook the longest journeys in the world. By the tenth century, however, shipping had specialized as boats operated between segments using intermediaries based in an increasing number of ports in Southeast Asia. This situation, Stargardt argues, played a pivotal role in a global trade system that stretched from Japan to the Mediterranean.[14] In China, the arrival of the Song dynasty in 960 led to an expansion in shipbuilding and a stronger inclination toward building trade ties with regional neighbors farther south. By the eleventh century, Arab merchants were returning to China's port cities, reestablishing connections of previous centuries, with Quanzhou emerging as the principal maritime trade hub along the country's east coast.[15] Examination of this period by Geoff Wade and Tansen Sen demonstrates the maritime economies that evolved under the Song (960–1279) and Yuan dynasties (1271–1368), eventually leading to the Ming-era (1368–1644) voyages of Zheng He.[16] It was during this time that Melaka entered its golden age of commerce, influencing the patterns by which commodities circulated within and across multiple regions. In his seminal work *Southeast Asia in the Age of Commerce,* Anthony Reid detailed the significance of Melaka in bringing together long-distance Indian, Chinese, Arab, and Persian merchants with their Javanese and Malay counterparts.[17] Sen has also subsequently argued that Melaka, along with Kochi and Malindi, were three port cities transformed by the Ming court into key regional trading hubs for the Indian Ocean region. More broadly, Zheng He's voyages enabled the Ming court to assert a strong influence over trade across the region. Accordingly, he states that "it was to acknowledge this naval supremacy, as well as to pursue the potential profits from commercial engagement with the Ming court, that polities from all regions of the Indian Ocean sent tributary missions to Ming China."[18]

In these various examples we see a series of historical processes that could receive significant attention from a language of Silk Road "revival." Belt and Road constitutes a platform for a rewriting of world history. As we saw at the end of chapter 3, the vision is for a "rising Asia" to place itself at the center of that history. There is a long-standing debate concerning the degree to which wealth, technologies, and ideas developed in Asia contributed to the ascendancy of Europe in the world system from the sixteenth century onward. Landmark studies by Chris Bayly, Kenneth Pomeranz, and Takeshi Hamashita have traced such connections between Europe and Asia, and have challenged assumptions concerning the economic and industrial development of Ming- and Qing-era China.[19] Sustained attention to the technologies and ideas that moved between Europe and Asia is likely to contribute new insights to such debates. A language of maritime and overland connectivity affords this, and the willing participation of many countries in such a program means Islamic, Persian, Southeast Asian, and East African histories are—potentially—all given new visibilities, new narratives. The faiths and traditions of the Middle East, for example, so often cast in the West as the source of conflict and intractable enmity, become expressions of rich civilizations and historical dialogue. Indeed, the trade and cultural exchanges of the overland Silk Road are central to S. Frederick Starr's thesis that Central Asia was home to a flourishing of cultural and scientific achievements spanning five centuries. Reworking the now-familiar notion of an Islamic golden age, Starr argues that the establishment of Baghdad as the capital of the Abbasid Caliphate in 750 heralded the beginning of an age of enlightenment, an era characterized by major achievements in art, architecture, science, literature, poetry, and philosophy. Two factors were critical to the flourishing and absorption of ideas: urbanization and widespread use of Arabic. Travelers to the region had long marveled at a highly urbanized Middle East and the great cities of the Central Asian heartland.[20] Among these, Starr venerates the grandeur and scale of Balkh, located in the north of present-day Afghanistan: "Balkh was by any measure one of the greatest cities of late antiquity. Its urban walls enclosed roughly a thousand acres, while the outermost walls that protected its suburban region and gardens were more than seventy-five miles in length. . . . [T]he citadel alone, called Bala Hisar, was twice the size of the entire lower city at Priene, a typical Hellenistic city on the Turkish coastline, and ten times the total area of ancient Troy."[21]

The city's prosperity stemmed from manufacturing; its successful cultivation of wheat, rice, and fruit crops; and the trade it developed with distant cities in India, China, the Middle East, and even the Mediterranean. Starr also points to numerous instances of flourishing art forms, documenting how architectural designs and concepts morphed through the incorporation of the iconography of different religions and their regional inflections. Domes offer a case in point, whereby design knowledge from Rome's Pantheon influenced mosque design in Turkmenistan and across the cities of Khorasan, with new ideas circulating back to Italy and influencing the grand buildings of Florence and Venice during the Renaissance.[22] More significant, however, Balkh constitutes a small part of the connected histories of urbanization across Central Asia, the Middle East, and the Indian subcontinent. As Christopher Beckwith argues, it is a history that demands greater recognition:

> The Renaissance occurred not only in Western Europe but throughout the Eurasian continent. In many respects it represents the artistic and intellectual apogee of Central Eurasia. While the European achievements in art, architecture, and music are well known, the achievements of the Islamic world, especially in Western Central Asia, Persia, and northern India, and of the Buddhist world, especially in Tibet, are much less well known. In the Islamic world, the Renaissance had begun at the time of Tamerlane, when Persian poetry attained perfection in the works of Hâfiẓ. Islamic miniature painting reached its height with the greatest Islamic miniature painter, Bihzad (ca. 1450/1460–ca. 1535), and others of the Timurid school of Herat. In 1522 Shâh Ismâ'îl, who patronized the arts in general, especially miniature painting and architecture, brought Bihzad from Herat to Tabriz.[23]

As the Silk Roads narrative valorizes and gives visibility to such themes, it alters how cultural heritage operates as a vehicle for nationalism and urban identities in the region. Indeed, one of the long-term implications of this emergent Silk Road heritage industry will be a corrective in how we think about urban history in the wider Asia region. Beckwith reminds us that the great metropolises and capital cities of the world were rarely located by the sea and instead have been inland, often linked to rivers. In European history, Paris, London, Madrid, Rome, and Athens substantiate this thesis.[24] Similarly, Tehran, Baghdad, Mecca, Isfahan, Angkor, Bagan, Beijing, Nara, Kyoto, and Delhi are among the great metropolises of Asia. Crucially, in the discourses of urban heritage that formed in the region in the second half of the twentieth century, and as exemplified in world heritage nomenclature, such inland cities have invariably been seen as "centers" or "seats" of culture, civilization, and power.[25] States, postcolonial or otherwise, have been the key drivers, ascribing value to their great cities as historical cornerstones of cultural, religious, or national identities. As Benedict Anderson and many others have argued, such inland settlements have been deployed to embody histories that build imagined communities for the modern era.[26] Angkor, Sukhothai, Hampi, Anuradhapura, Kathmandu, and Isfahan have all become pivotal in ethno-cultural nationalisms. Their designation as modern heritage sites by international agencies such as UNESCO has reinforced their centripetal power, spaces around which nation building and postcolonial identities have revolved. Even for those iconic settlements located on the overland Silk Road, their significance has often been designated in terms of their influence as centers and heights of cultural or religious life. The World Heritage listings of Bukhara and the Mausoleum of Khoja Ahmed Yasawi in Uzbekistan and Kazakhstan, respectively, are illustrative of this issue. The statements of significance for inscribing Bukhara indicated the town "was the largest centre for Muslim theology, particularly on Sufism . . . in terms of its urban layout and buildings had a profound influence on the evolution and planning of towns in a wide region of Central Asia."[27] In the case of the mausoleum located in the town of Yasi, UNESCO recognized the site as "an outstanding achievement in the Timurid architecture . . . the mausoleum and its property represent an exceptional testimony to the culture of the Central Asian region."[28] Listed in 1993 and 2003, respectively, these nominations present culture and religion in the singular and define value in terms of the "exceptional" and "profound." This lies in significant contrast to the language of "crossroads" and "meeting points" in the speeches offered in support of similar sites nominated to the World Heritage List in recent years.

The world heritage movement has long been criticized for overemphasizing Europe in its representation of world history and the material legacy of different civilizations, and it may well be argued that such Eurocentrism has been at play here. In relation to the geographical scope of this book, the commemoration of premodern forms of globalization and long-distance trade within discourses of heritage have primarily been tied to forms of maritime connectivity and the port cities that emerged in the fifteenth century, when contact between Europe and Asia grew rapidly through the trade relations established by Portuguese, Dutch, British, and French.[29] In narrating the global histories of connection, flows, and exchange across Asia, contemporary heritage discourses have privileged an era of European "discovery," colonialism, and the intraregional maritime trade of such periods. The World Heritage Site listings of Penang, Galle, Macau, Hoi An, and Old Goa exemplify this pattern, all celebrated for their outward-looking, "cosmopolitan" pasts. The notion of a Maritime Silk Road pushes the story further back and traces a multitude of intraregional connections that predate the European "age of discovery." My argument here rests on the construction of some prototypical categories and a recognition that Xi'an and Samarkand as inland cosmopolitan cities complicate any simple analysis (fig. 6.1). But what I am pointing to are some overarching differences in the ways in which inland and coastal sites have been ascribed their value and historic significance within the international registry established by UNESCO. The recent development of multicountry nomination collaborations for the overland Silk Road routes begins to address this imbalance, in that it shifts the narrative of inland urban areas away from a language of centers and the unique and extraordinary toward a discourse of hubs and contact between cultures and civilizations stretching over great distances and hundreds of years. I do not anticipate this will simply overwrite existing framings of cities, erasing their symbolic values as national sites. Chapter 2 identified some key characteristics of cultural governance in the Middle East and Soviet Union, and their legacies continue to play out today.

6.1 Walled city of Xi'an. Courtesy of George Oze / Alamy.

Indeed, across Eurasia, museums, historical sites, and various fields of scholarship have long formed part of the state apparatus for constructing ethno-cultural and ethno-religious nationalisms.[30] National museums in Almaty, Tehran, Tashkent, Ankara, and elsewhere continue to adopt a museology that ties ethnographic exhibits to natural history and displays of technology in the construction of a teleological narrative of the nation. For a number of countries across the region, modern historiography has also been oriented around the periodization of empires and kingdoms and their associated territories. Mozaffari illustrates this in the making of a modern Iran and the sense of homeland created through the mythical narratives of the Achaemenid Empire and later Safavid dynasty.[31] But as Hamid Dabashi and Minoo Moallem both note, it was the guilds and bazaars, and the objects of trade such as carpets, that incorporated Iran into the regional and global economy.[32] The focus on Silk Road objects in the previous chapter can thus be extended to tourist crafts and likely increases in the export of handcrafted items associated with the Silk Road story. In essence, then, any shift to a geography of regions built around a paradigm of connections will no doubt be slow and inevitably partial. The Silk Road is likely to bring new infusions to existing frameworks, such that different inflections of meaning and value are emphasized in different contexts and at different moments in time, and the national and international are continually entangled.

In acknowledging the various structures and norms that work to resist change, the conjecture here is that as an arena of heritage production, the concept of the Silk Roads, as advanced through Belt and Road, will go some way to making the cultural paraphernalia of national identity less inwardly territorial, the identities of the region's cities less culturally centripetal, and, perhaps, the narratives of world history a little less Eurocentric. This is certainly not to say that the competing forces that lie at the heart of discourses of history and heritage in the region will be smoothed out and realigned into a singular metanarrative. Even if we take just the processes and trends under consideration here, it is clear some distinct fault lines and hierarchies will remain. With China as the lead author in much of this, we are seeing the solidification of a Sinocentric version of Eurasian interconnectivity through a system of cooperation characterized by asymmetries in size, resources, and power. The themes highlighted so far point to the fragmented and multicentered trade networks that preceded the routes now being constructed for the Maritime Silk Road of the twenty-first century. Beijing's vision of reviving these seaborne and coastal histories is being diplomatically buttressed by the seven voyages of Zheng He, shipwrecks lying in the waters of Southeast Asia, and submissions to the World Heritage Center for Maritime Silk Road listings. Such developments speak to China's desire for greater international recognition of its civilization and its impact on other societies, near and far. Clearly, the heritage diplomacy of Belt and Road advances this, forging narratives of pioneering adventure and heroism that obscure others. Little space, for example, is given to the fact that Zheng He's routes were well charted by earlier Arab and Persian sailors and that his fleets benefited from the trade networks established within an Islamic cosmopolitan world stretching from Indonesia and Malaysia in Southeast Asia, to India and East Africa. Given that a number of the fleets involved more than three hundred ships, Geoff Wade and Tansen Sen are among the historians arguing that his voyages were designed to cement the hierarchies of a Sinocentric tributary system by powerfully conveying China's technological and political strength to periphery countries. The deployment of tens of thousands of well-trained soldiers points to military and expansionist desires held by the Ming Empire.[33] Wade depicts various regional conflicts to argue the voyages need to be read as violent adventures of an imperialist polity.[34] Citing excerpts from the *Ming Shilu*, the imperial annals of the Ming dynasty, he documents a major attack in 1407 on Sumatra, which led to five thousand deaths and the burning and capturing of ships. The same year 170 of Zheng He's troops were killed as they fought with local forces in Java in an attempt to gain control of the local trade structures.[35] For both Sen and Wade, though, his third visit to Sri Lanka of 1410 is of particular significance. A military battle led to physical damage inflicted upon the royal capital and the installation of a puppet ruler with the existing king carried back to the court in Nanjing.[36] As we have seen in previous chapters, such inconvenient histories are smoothed over in today's narration of Zheng He's voyages, as Sri Lanka and China seek to strengthen their bilateral relations. The reconstruction of the memory of Zheng He's visits to the island form part of a disposition for "conflict avoidance" in a Silk Road history and heritage. Within the overarching story of peaceful exchange and civilizational enrichment spanning centuries, battles and bloodshed are an inconvenience eagerly passed over. The military campaigns of Genghis Khan, the extended wars the Chinese Tang dynasty conducted with the Arab Abbasid caliphate in the eighth century, and the slaughter of tens of thousands of Arab and

Persian merchants in the Yangzhou and Guangzhou massacres of 760 and 878, respectively, are among the many historical events rendered invisible in a Silk Road narrative fueled by nostalgia and pursuit for convenient truths.[37]

In this art of forgetting, tourism plays a critical role. The very nature of tourism is to rewrite the past in ways that render it suitable for public consumption. But ongoing shifts in the demographics and infrastructure of Asia's tourism markets will contour where and how the stories of the Silk Road are told. The corridors of world heritage tourism now in development through northern China and Central Asia mean that the complex connections and exchanges that took place across vast mountain ranges and plateaus will be funneled through those locations of antiquity that are accessible and amenable to a Silk Road tourism industry. The Chinese tourist, now in long-term ascendancy, will also increasingly provide the default reference point for building histories of long-distance connection. As exhibits of porcelain in museums and Zheng He attractions as far apart as Kenya and Indonesia capture the imagination of their Chinese visitors, a Sinocentric version of the Maritime Silk Road takes hold.

The degree to which all this opens up space for other stories depicting how religions, technologies, crafts, or ideas traveled between locations across the Indian Ocean and beyond remains unclear. Both East Africa and Indonesia have long been ignored in the historicization of the Indian Ocean.[38] The story of Zheng He's voyages begins to connect such locations. But as the fascinating study by Thomas Vernet indicates, Swahili merchants regularly plied the waters of the west Indian Ocean and the Arabian Sea in the early modern era to reach Muscat, Oman, the coastal ports of Yemen, and across to the west Indian states of Gujarat and Goa (fig. 6.2).[39] The insights produced by scholars working on such themes also shows China's impact on the region stretches back only so far. It is thus evident Belt and Road only calls on certain geographies and times to be revived, largely ignoring others, in some cases deeper, histories of maritime connection. Naturally, parallel critiques also need to be offered for the contours of connectivity being fashioned into shape through the heritage diplomacies of the overland Silk Road. Anticipating how such relations will play out is beyond the scope of this volume. However, the final section of this chapter turns to a development that is influencing these unfolding futures in important and fascinating ways, that of how expertise across a number of scholarly disciplines is being brought into new configurations.

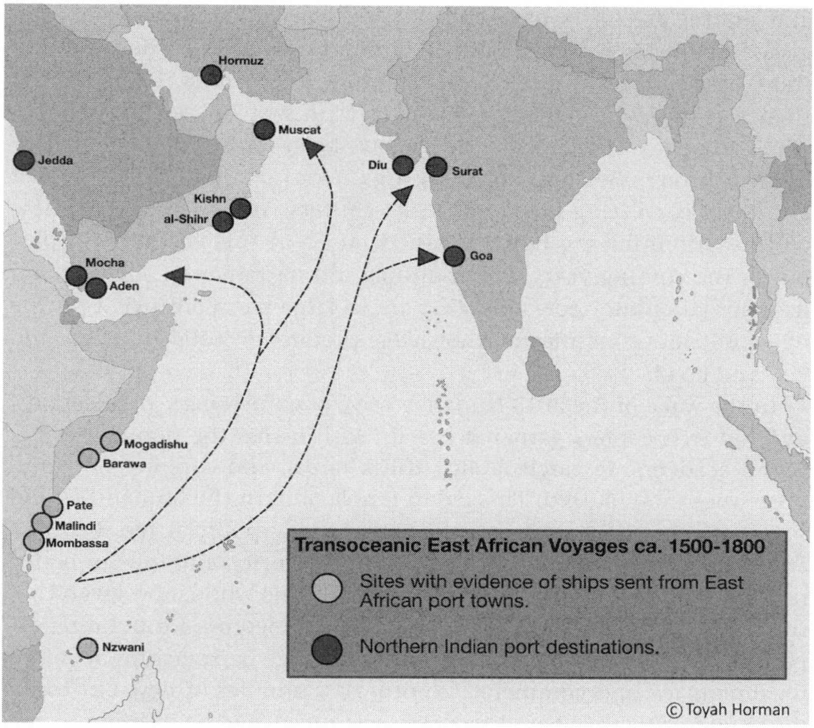

6.2 West Indian Ocean and Arabian Sea routes of Swahili merchants. Courtesy of Toyah Horman.

Assembling Belt and Road Knowledge

This organization aims at restoring "Silkroadia," the Silk Road spirit—a symbol of the bridge between the East and West by banding together universities located on the land and sea routes of the Silk Road and contributing to world peace and the creative development of civilization by training future leaders devoted to this spirit.

CONSTITUTION, ARTICLE 2.1, SILK ROADS UNIVERSITY NETWORK[40]

It comes as little surprise that research and the information produced about the South China Sea is a highly politicized affair. The disciplines of law, international relations, and defense studies have formed the backbone of research institutes and think tanks across the Asia-Pacific, with experts often strategically located close to national centers of power.[41] To establish knowledge certainties around territories and jurisdictions, political science has converged with military studies, and marine scientists have collaborated with underwater archaeologists.[42] Zheng He has even been the subject of articles produced by authors based in naval war colleges in the United States.[43] My interest, then, lies in laying out some of the ways in which academic scholarship and policy-oriented knowledge have been transformed and reconfigured through Belt and Road. My argument is twofold: first, that Belt and Road is creating new knowledge clusters, and second, that these will heavily influence how the issues raised in this chapter will play out in the coming years. The framings and narratives of the past and its material culture now emerging are in large part contingent on the new political economies of knowledge production solidifying around Belt and Road.

In the wake of the 2013 launch, a wave of conferences, publications, and research centers gathered speed. Within this the boundaries between academic research, policy think tanks, and commercial enterprise were soon blurred. This led to scholarship in the humanities and social sciences being drawn into entirely new political terrains. Belt and Road affects the energy, construction, security, and foreign-policy portfolios of dozens of governments around the world, and given that the think-tank template is a well-established medium for building expert advisory capacities and the international partnerships required for diplomacy and commercial ventures, a number of new BRI think tanks and university-based research institutes were established in the first three years (appendix 1). Many of these are multidisciplinary, with the majority orienting their intelligence-building capacities toward the cooperation pillars of the 2015 Visions and Action Plan. It is a space that many universities have been eager to play in, often by pooling existing expertise in the effort to secure funding, government or otherwise, and build networks for international student recruitment. In this regard, geography plays a critical role. In the case of the Collaborative Innovation Center of Silk Road Economic Belt Research, at Xi'an Jiaotong University, for example, the priority is to build ties with a number of Central Asian countries, as well as with Russia and India. In proposing collaborations across multiple fields—law, economics, engineering, trade—the center launched with the ambition of contributing to the diplomatic dimensions of BRI by building an "academic belt of the Silk Road" that would improve "exchanges between experts from different countries and their ability of knowledge production, tackling key problems by a collaborative innovation which was transnational, transregional and transboundary."[44] It was a template for "win-win cooperation" that included nurturing a "dialog among civilisations" as one of its priorities.[45] For those universities nearer the coast, the assemblages look a little different. In Xiamen, Huaqiao University's Institute of Maritime Silk Road is using its location on the Taiwan Strait and the university's historical connections with overseas Chinese to build a series of partnerships, including ones focused on "cultural exchanges" with countries across Southeast Asia.[46] Farther south, City University of Macau established the Macau One Belt One Road Research Center, with a particular focus on strengthening connections between China and Portuguese-speaking BRI countries.[47] Maritime histories linking Portugal, Timor-Leste, Mozambique, and Macau underpin a networking strategy revolving around the twenty-first-century Maritime Silk Road. With Belt and Road holding profound consequences for Hong Kong, the city witnessed a flurry of Belt and Road academic production, including institutes created across a number of universities. Portfolios of expertise built around trade, banking, and the shipping and air transport sectors featured research projects on Hong Kong's place in the historic Maritime Silk Road to develop a Belt and Road rhetoric of public diplomacy and international cooperation.[48]

Belt and Road has also become a vehicle for Chinese-funded research centers in neighboring countries. In 2016, the Royal University of Phnom Penh created such an initiative, the Maritime Silk Road Research Center. For the Cambodian government, a long-term recipient of Chinese aid and investment, the center provided a gateway for sourcing further funds from the Asian Investment Infrastructure Bank and its Silk Road Fund.[49] In June 2015, the Confucius Institute of Maritime Silk Road was inaugurated as a collaboration of twenty-seven educational institutions across Thailand. Designed to be a model for similarly named entities in other BRI countries, the institute's key remit was to build educational and cultural exchanges between Thailand and China.[50]

Such educational bilateralism has also extended into various multilateral configurations. These include the Silk Road Think Tank Network, the New Silk Road Law Schools Alliance, Maritime Continental Silk Road Cities Alliance, Belt and Road Think Tank Association, University Alliance of the New Silk Road, and the International Silk Road Think Tank Association. In November 2014, the Silk Road Universities Network was established with the premise of fostering academic connections that revive the "Silkroadia" spirit of the ancient Silk Roads, a commitment laid out in the organization's constitution and mission statement:

> We aim to realize our vision of serving the world by undertaking diverse projects that restore the historical value of the ancient Silk Road which has been a source of immense pride for centuries as the birth place of four major civilizations. The most valuable lesson from the history of the Silk Road is that the key to peaceful coexistence and collective prosperity is to treat individual differences as a cause for celebration rather than segregation, best captured in Silkroadia-the spirit of ancient Silk Road. In line with this thinking, we believe that the coming together of universities can help realize this vision through fostering an exchange of ideas culminating into decisive action between intellectuals transcending national, religious and cultural boundaries.[51]

To this end, annual conferences provided the focal point for the development of the associated International Association for Silk Road Studies, involving scholars from universities in more than twenty countries, including Iran, Malaysia, Pakistan, Sri Lanka, Portugal, and Greece. Not surprisingly, the number of academic and policy conferences dedicated to Belt and Road also expanded rapidly in the aftermath of its launch. Between 2014 and 2016 an extraordinary array of One Belt One Road–themed conferences took place across the world. Clearly, it rapidly became a container for branding existing areas of inquiry. But in a significant number of instances, the geographic and economic possibilities of Belt and Road afforded new configurations of knowledge production. Political scientists have been brought into dialogue with anthropologists and archaeologists. Indeed, for those working across the humanities and social sciences, Belt and Road has further increased the gravitational pull China exerts on debates about development and globalization, neoliberalism and neocolonialism. It has also created new dynamics between nationalism and internationalism, with maritime archaeology serving as an interesting example. As we saw in previous chapters, a number of maritime archaeology research centers have been established across Asia in the past two decades. State funding has meant that national and territorial interests have quietly underpinned the parameters of maritime exploration and the salvaging of artifacts. This has been clearly reflected in the delivery and production of academic papers for the field.[52] Given the intense sensitivities concerning archaeological discoveries in the South China Sea, shifting to a language of "win-win cooperation" is far more complex than it is for collaborations over serial world-heritage nominations for overland Silk Road properties.

Interestingly, the Maritime Silk Road of the twenty-first century brings Hong Kong and Singapore into this picture, with both having much a greater investment in the development of maritime heritage sectors than they did previously. Indeed, both cities reveal the confluence of scholarly research on Silk Road histories, Belt and Road futures, and governmental interests. As part of a wider BRI public diplomacy program, the government of Hong Kong instructed its public-sector institutions to develop initiatives addressing the city's future in the Maritime Silk Road. In response, the city's art and history museums ran events exploring the cultural and historical roots of Silk Road cooperation.[53] During the 2016 conference, the Belt and Road Initiative—Combining Hard with Soft Power, maritime archaeology was among the topics debated as an asset for trade and diplomacy.[54] And in late 2017, the city's maritime museum contributed to the hosting of the triennial Asia-Pacific Regional Conference on Underwater Cultural Heritage. In Singapore, academic activity around maritime histories also accelerated. Topics for university-based conferences included mercantile connections established under Srivijaya and between China-Arabia encounters.[55] The previous chapter highlighted how both cities have hosted a number of maritime heritage exhibitions, and in each case the personal connections between experts in academia and the public museum and heritage sectors are well established. Scholars in history, archaeology, and anthropology are regularly called on to offer guidance on exhibition planning. Equally, academics frequently look to cultural-sector public institutions for event hosting and collaborative funding. Singapore and Hong Kong are also among a select group of cities in Asia whose universities have the funding and international connections required for hosting large-scale scholarly events and conducting high-impact research projects. Of course, the key player here is China, and the collaboration between Tsinghua University and University of Athens noted earlier constitutes a small example of the new historical insights being produced on the back of Belt and Road. In contrast, a number of middle-tier countries across the region—Iran, Vietnam, Bangladesh, and others—lack the capacity to undertake expensive underwater-archaeology research initiatives. Crucially, and as we have seen in the case of Sri Lanka and Kenya, this means importing resources and expertise, a process that strongly determines the nature of work undertaken and how the flow through from research to the production of knowledge and subsequent discourses of public heritage occurs. My point, then, is that where archaeology starts, heritage frequently follows. Singapore and Hong Kong are good examples of where the threads of connection between scholarship and public discourse can be clearly found. But as cities across the region look to host exhibitions, develop archaeological sites, and build heritage-tourism attractions, scholarly expertise on Silk Road histories and connections becomes increasingly important.

There are also instances of the feedback loop beginning to operate in reverse, where extant discourses of heritage inform academic knowledge production. A collaboration between Nanyang Technological University and the International Zheng He Society in 2015 is illustrative of this point. A conference became a book, titled *China's One Belt One Road Initiative*, published in London a year later.[56] The premise of the project was to understand the "central OBOR concept of connectivity,"[57] and with a particular focus on Southeast Asia the book's arguments are buttressed by lengthy accounts of trade and vassal kingdoms of the tenth to fourteenth centuries. To this end, Tseng Hui-Yi states: "The entire world system functioned smoothly when the connection through China operated well. Actually, this was the original apparatus upon which the 21st Century OBOR initiative was sketched out."[58]

The volume further cements these temporal and spatial connections through assertions that the twenty-first-century "maritime route is almost identical with Zheng He's seven maritime voyages."[59] Indeed, Tan Ta Sen, the president of the International Zheng He Society and owner of the museum in Melaka, argues that Zheng He's "Art of Collaboration" holds important insights for building cordial relations today:

> Cheng Ho's voyages to the Western Ocean showed such features of Confucian ethics as humanism, benevolence, righteousness, forgiveness, morality, harmonious interpersonal relationships, as well as a stable social order. He incorporated Confucian ethics into his foreign policy as a form of the "Art of Collaboration." . . . The Cheng Ho spirit could be a model for contemporary international relations and foreign relations. Reliving Cheng Ho's spirit in international relations based on mutual respect, non-invasion, non-intervention and fostering good relationships with foreign states will result in the creation of a world order where multipolar powers are in a partnership to achieve world peace, universal harmony and equality.[60]

In drawing these various examples together, we see the academic entrepreneurialism that has formed around Belt and Road. Some institutes and networks will no doubt be far more successful than others over the longer term. What we also see in the examples of Hong Kong, Singapore, Xiamen, and Xi'an is the incorporation of universities into government strategies for making cities economically competitive. More broadly, Belt and Road significantly advances the neoliberal agenda that frames higher education in terms of national priorities, rearranging scholarship into new configurations, both intellectually and institutionally. It is an arena that involves Silk Road histories and cultural pasts being invoked in multiple ways. In certain contexts, this idea of building contemporary trade and business relations on the foundations of links established more than half a millennia ago is being critically interrogated. But as we have seen here, the notion of ancient overland and sea-based connections is also being deployed as a vehicle for universities to conduct public diplomacy, recruit students, undertake cultural exchanges, and build networks of high-value commercial and scientific research. Once again, then, we see a political economy at play shaping how the past is conceived, molding which histories are brought into focus, and how they come to be funneled through a language of exchanges and partnerships. In this final example from Singapore, we also see how certain interpretations of Southeast Asia's maritime history give intellectual weight to analyses that advance the government's ambitions of ensuring the country remains tightly integrated in Asia's networked economy.

Belt and Road has proved an extraordinary catalyst of knowledge production. From a starting point of twenty-two in 2014, a Google Scholar search returns more than one thousand "One Belt One Road" articles published in English two years later.[61] The majority of these fell within the domain of political geography, foreign affairs, and business. Belt and Road research themes and topics will continue to diversify and proliferate. The aim here has been to point to some of the institutional and political forces that will continue to mold the complex histories of Eurasia and the Indian Ocean into particular narratives and configurations. An overriding concern for recognizing, reviving, and building connectivities opens up important spaces for understanding previously neglected historical events and processes. But as we have seen over the course of this chapter, the forms of reterritorialization that Belt and Road delivers steers these histories along certain intellectual and geographic corridors and across institutional networks and hubs.

SEVEN

Geocultural Power

The Routes of Geocultural Power

As the conflict in Syria entered its fifth year in 2016, Russia increased its support for the government of Bashar al-Assad. Russian air strikes on Aleppo that killed hundreds of civilians were widely condemned, with French president François Hollande among those responding with the proposal of war-crimes tribunals. Russian president Vladimir Putin had long justified his intervention in Syria as a fight against the Islamic State and terrorism. But in an increasingly tense and hostile diplomatic environment, various observers expressed concern that the conflict could escalate to a wider regional, or even global, war as negotiations between Moscow and Washington broke down. In March 2016 Russian troops took control of Palmyra from IS. Four months later Putin used the city's archaeological remains for an extraordinary spectacle that brought together the three themes that had dominated international media coverage since the conflict began—the humanitarian crisis, destruction of antiquities, and the military conflict itself. The St. Petersburg–based Mariinsky Theatre's orchestra performed Bach and Prokofiev in Palmyra's Roman amphitheater to an audience of journalists, UNESCO ambassadors, military and religious leaders, and members of the Russian culture ministry (fig. 7.1).[1] Following an opening speech by the conductor Valery Gergiev, Putin delivered a message by video from Moscow indicating that the concert, *A Prayer for Palmyra,* was offered as an

7.1 Mariinsky Theater Orchestra performance in Palmyra's Roman amphitheater, July 2016. Courtesy of EPA European Pressphoto Agency / Alamy.

act of remembrance and hope. With the event including a large picture mounted on the stage of Khaled al-Asaad—the archaeologist executed for refusing to reveal the whereabouts of hidden antiquities—the contrast to the violence of previous months was dramatic, particularly given that IS had filmed and broadcast the execution of twenty-five prisoners at the amphitheater just over a year earlier.[2]

For many observers in the West, such "orchestrated diplomacy" was little more than a cynically conceived spectacle. That aside, the performance represented an explicit attempt to claim some moral high ground in the bloody conflict and propaganda war. In his message, Vladimir Putin expressed "hope that our contemporary civilization will be relieved from the horrible disease of international terrorism."[3] The transposition of Russian classical music onto one of the Middle East's most important landscapes of antiquity delivered a powerful image of Moscow leading the fight for Western civilization. Symbolically, it placed Russia on the opposite side to the cultural barbarians and their ideologies of destruction, and it suggested that by liberating the Silk Road city of Palmyra, Russian troops were the guardians of high culture for the world.

The themes addressed in this book put forward a case for looking beyond the geopolitical and conventional notions of diplomacy for understanding international affairs. I have explored the analytically elusive concept of culture and the expediency of well-crafted, stylized pasts. To provide some analytical and historical purchase on these, chapter 2 presented a biography of the Silk Road as a geocultural form of the modern era. As others have suggested, the geocultural opens up themes and lines of inquiry unexplored by geopolitics, and yet it remains largely underdeveloped as a concept. In 1991 Immanuel Wallerstein made the case for seeing it as akin to the geopolitical. However, in an account that pivoted on the idea that geoculture can help explain core-periphery shifts in the modern world system, particularly in the late twentieth century, Wallerstein offered little clarification as to how the term might be conceptualized.[4] More illuminating is Ulf Hannerz's discussion of geocultural scenarios. As Hannerz reminds us, the culture concept is inherently flexible in terms of scale. By implication, then, culture comes to be imagined at multiple scales, and by inserting the *geo*, we can critically interpret the different mapmaking processes, which seek to make sense of "the distribution of things cultural, somehow cultural, over territories and their human populations."[5] Undoubtedly, the nation-state has become the most entrenched geocultural imaginary of the modern era. But the Great Game, the Cold War, and the Silk Road itself are among those geocultural categories that give form to larger, more amorphous entities. As Hannerz suggests, they help us "*think* geoculturally, about the world and its parts, and the main features of those parts."[6] So although the merits of categories such as the Far East or Third World have dwindled in the past few decades, and rightfully so, the arguments laid out here indicate that the Silk Road is a geocultural form in ascendancy.

A biography of the Silk Road as connectivity indicates how China is now using this geocultural form for its own ends through the Belt and Road framework, a situation that sustains the intersections of culture, geopolitics, and infrastructure first established in the late nineteenth century. As the "home" of silk production, China is able to insert itself at the center, both culturally and geographically, of a story of regional and East-West contact. Various chapters here have highlighted how the language and technologies of heritage communicate this to Belt and Road partners. Expressed in a variety of material and nonmaterial forms, the story of the Silk Road fashions slices of Chinese history into a geocultural resource, one that is doing considerable political work in the heritage diplomacy of Belt and Road. In their plural, the Silk Roads act as a platform for demonstrating the global impact of Chinese civilization. Undoubtedly, this is intended to reach international audiences. But the promotion of the story domestically is also about nurturing a form of Chinese citizenry that is anchored in an holistically conceived vision of a deep culture that has benefited from the integration of outside influences whilst successfully retaining a strong sense of integrity and distinctiveness. As one of the proponents of such a vision for China, Zhang Weiwei notes that "a civilizational state has exceedingly strong historical and cultural traditions [and] has a strong capability to draw on the strengths of other nations while maintaining its own identity."[7] As we have seen, testimony to this "heritage" can be found both at home and abroad. As outbound travel continues to grow, Chinese tourists are learning about the spread and impact of their culture around the world. The ceramics presented in figures 5.2 and 5.3 form part of the Sir Percival David Collection in the British Museum in London. On any given weekday, soon after opening, the room fills with groups of Chinese tourists admiring and photographing the wares on display. The story of the Silk Road puts these objects in an all-expansive narrative of Chinese global history. Engaging with the intense debate that has emerged in the past two decades concerning the task of defining and rejuvenating Chinese culture and civilization for the modern era is beyond the scope of this discussion. However, to briefly connect the past to the present through the concept of the civilizational state, Randall Collins's discussion of civilizations as "zones of prestige" helps pinpoint some key dynamics in play here. For Collins, civilizations exert a power of attraction that radiates outward via unevenly formed networks and along particular channels. Offering some historical examples, he identifies sojourners and pilgrims, traders and students, as among the agents of prestige transmission, whether they be traveling inward toward centers of creativity and activity or sent out to cross great distances and border zones to communicate particular ideas and values. Centers of Buddhist pilgrimage and learning in India, as well as missionaries and teachers migrating between China and India, are cited as examples of such processes. Naturally, the influence of civilizations flourishes and fades, with attraction being generated "through symbolic objects, some of them verbal and literary, others artistic and embodied in physical artifacts that people travel to see. We can study civilizations sociologically by studying the activities that raise and lower the level of civilizational charisma and the degree and distance of its attraction."[8]

Crucially, on the back of such cultural contact, economic and political structures appear, and, as Collins notes, the degree to which absorption and reproduction occurs is contingent on the inequalities of power in the contact process. Collins's analysis thus indicates the significance of "revisiting" and "reviving" the Silk Road as a geography of prestige today.

Chapter 2 demonstrated that the international circulation of the Silk Road concept is contingent on a particular ecology. In this regard, Belt and Road is transformative, creating new gravitational forces within its sweeping geographies. As noted earlier, such developments are fostering anxieties and countermoves by others in the region. To put this situation in a larger context, we can turn to Alfred Rieber's framing of Eurasia as a history of competing geocultural spaces. Drawing inspiration from the Annales school tradition of historiography and its concern for longer-term regional transitions, Rieber maps out a past of imperial rivalry and struggle between multicultural states and groups resisting subjugation and assimilation. A key factor in Eurasia's history, he argues, is the large-scale movement of populations. "Population shatter zones" were thus the frontiers of violence and subversion, as Mongol, Qing, Ottoman, Qajar, Russian, and other empires sought power, legitimacy, and territory.[9] For Rieber, these deeper historical patterns offer guidance for interpreting events of the twentieth century. As he states, "By treating Eurasia as a contested geocultural space, Russian expansion is placed in a different context, as a product of centuries old struggle among rival imperial powers."[10] To what degree, then, does Rieber's thesis reveal underlying patterns in China's geocultural imagining of Eurasia as the Silk Roads today?

Clearly, the strategies for acquiring influence and resources in the twenty-first century are markedly different from those of the nineteenth. But if securing new territories has (largely) been consigned to the annals of military history, the significance of transboundary infrastructure and international trade has been only elevated. In the political map of today, power is accumulated by building or being part of these international networks. Parag Khanna's neologism of *connectography* seeks to capture this landscape of international cooperation and competition, asymmetry and great-power rivalry. The themes explored here bring cultural connectivities into this frame. People-to-people ties might be about building trust and dialogue cross-culturally, but they are also about reaching populations with cultural or religious affinities. China, India, and Russia remain multicultural states with great power ambitions. In all three, twentieth-century practices of statecraft involved forging histories and identities around the geobody of the nation. This is not in abeyance or decline today, but what we see in the Silk Road is the appropriation of a narrative of the past that is both nationalist and internationalist, and, crucially, not claimed as an exclusive patrimony. Unlike Persepolis and the Achaemenid Empire or Ayutthaya and the Kingdom of Siam, the Silk Road does not proclaim the kind of etymological or territorial roots around which modern states would "naturally" craft their cultural nationalisms and paraphernalia of heritage. Even in the case of China, which claims to be the home of sericulture, the story of the Silk Road enables silk to be presented as its gift to the world.

Back in 1989 Bruce Trigger suggested that, from its birth, archaeology has been the handmaiden of nationalism, colonialism, and imperialism.[11] But as the latter have been decoupled from territorial acquisition and the formal governance of populations, analyzing the role of material culture histories in such processes becomes more complicated. The construction of Sinocentric narratives of connectivity points to expansionist ambitions. Indeed, William Callahan has argued that the emphasis rising powers place on connectivity today forms part of their larger strategies for building regional, and perhaps even global, order based on new ideas and rules about governance and cooperation. Beijing's notion of connectivity extends far beyond hardware to include the linking up of "ideas, institutions, and behavior in diplomacy itself."[12] If this is the case, the Silk Roads of the twenty-first century represent the key strategic framework for remaking the international order in ways that shift political and economic power toward a China-centric Asia. Further insight into how this might work comes from Timo Kivimäki and Antara Ghosal Singh, who make the case that the version of soft power and international diplomacy China is advancing qualitatively differs from that of Western countries and the Cold War era.[13] Kivimäki argues the definitions of soft power as attraction, advanced by Nye, Shambaugh, and others, inadequately accounts for changes in the international system since the 1990s and why China's noninterference policy means that distinctions between hard and soft power are far less well defined.[14] Win-win economic policies and cooperation over highways and energy infrastructure are soft uses of hard power that produce attraction but do not necessarily involve the forms of coercion or alignment familiar to the tussles of the Cold War. Rethinking definitions of soft power to incorporate different modes of connectivity, aid, infrastructure, and shared interest suggests China has been far more successful in securing its influence around the world than analysts who use the experience of the United States as a yardstick have suggested. For Callahan, then, what we are seeing is an interweaving of contradictory elements, whereby nationalist insecurities drive both national security and foreign policy, a dynamic that produces a mutual entangling of soft and hard power on the international stage.[15]

The Silk Road metaphors of peace, trust, friendship, and harmony we saw in chapter 3 speak to these complex dynamics in significant ways. With a relative deficit of hard power, Beijing has sought influence by positioning itself as a civilizational state, one that holds qualities and values for global security and stability. In marked contrast to the "clash of civilizations" between East and West, as framed by Huntington and his followers, China invokes civilization as a resource for peace, respect, and harmony: "Civilizations are like water, moistening everything silently. We need to encourage different civilizations to respect each other and live together in harmony while promoting their exchanges and mutual learning as a bridge of friendship among peoples, a driving force behind human society, and a strong bond for world peace. We should seek wisdom and nourishment from various civilizations to provide support and consolation for people's mind [*sic*], and work together to tackle the challenges facing mankind."[16]

In January 2016 Xi Jinping also told the League of Arab States in Cairo that "Chinese and Arabian civilizations share common ideas and pursuits accumulated in mankind's development and progress."[17] But as we saw in the examples in chapter 3, humiliation—and China's memories of a century of rebellions, conflicts, and invasions—continues to cast its shadow. Xi's pronouncements to audiences at the G20 and World Economic Forum in Davos that Eastern traditions offer a bedrock on which a new global economic order can be built reflect a desire in China among those who wish to see the country regain its rightful standing in the world.[18] An implicit component of the Silk Road "spirit," then, is the sense that it constitutes a geography of resilience and renaissance.[19] China has promoted the Silk Roads as a shared reclaiming of dignity and sovereignty, a new flowering after histories of humiliation. Here we can return to questions of geography, as discussed in chapter 2. There it was noted that within Asia, histories of intraregional connectivity have been central to the conceptualization of the Silk Roads. In the early twentieth century the political impetus for this was a resistance to European hegemony. With the idea of a twenty-first-century Silk Road as common history, shared heritage continues this tradition, and indeed carries the specter of the "Asian values" discourse. Commentators on Belt and Road in the West have normatively understood the Silk Road as a "revival" of East-West connections. Such analyses are reaffirmed in infrastructure tie-ups to Greece, Germany, and beyond, but the cultural dimensions of BRI point toward a more exclusionary discourse. After all, as Yang Jiechi noted in chapter 3, the revival of the Silk Roads are the coordinates by which we can track a resurgent Asia, with the Middle Kingdom at its core.

Such themes resonate with interpretations of Chinese leadership in international affairs in relation to *tianxia*, or "all-under-Heaven." A traditional concept predating Confucius, *tianxia* has received renewed attention through the writings of Zhao Tingyang, Yan Xuetong, Howard French, and others.[20] Offered as an alternative to an international system oriented around antagonistic nation-states, *tianxia* ties legitimacy in political leadership to what Ban Wang summarizes as "a sphere of culture and value—a set of ideals and conduct to be internalized by all individuals and groups."[21] For Wang and Prasenjit Duara, this involves building leadership and a welfare for humanity around the ideals of aesthetics and cosmopolitanism.[22] Hierarchy and asymmetries of power are readily acknowledged, with China, as a great power, charged with providing moral and political leadership. For critics such as William Callahan, such a framework legitimizes forms of cultural and political hegemony, and a world oriented around *Pax Sinica*. As Howard French reminds us, while *tianxia* primarily pertained to the known world of Asia and the shifting power base of the Han (Chinese) vis-à-vis the Turkic, Mongol, Manchurian, and others, today the cosmology of empire is global, with the United States being the primary obstacle to a twenty-first-century Sinic world order.[23] Geoff Wade's analysis of the Ming chronicles also reveals how the "civilizational rhetoric" underpinning this hierarchy of interpolity relations exalted peace and ideas of virtuous leadership. But as he goes on to show, in its deployment this rhetoric was in part "utilized to present the motives and actions of the emperor and the Chinese state in ideological terms that concealed from both foreign and, more important, Chinese audiences the true nature of China's foreign policies and practices."[24] Stevan Harrell and Edward Vickers are among those who have also demonstrated the longstanding presence of a "civilizing mission" within the Chinese state, both at home and abroad.[25] This begs the question of how we interpret the civilizational rhetoric of today.

Given that the political landscape of Eurasia has changed profoundly, and any previous formations of sovereignty and trade have been overlaid with a nation-state structure, simple analogies belie the complexities of current events. We are seeing China, however, as a civilizational state, exercising its geocultural advantage through the Silk Roads at a time of increasing regional and global competition. More specifically, constructed as a heritage of connectivity, the Silk Roads directly align with the wider connective agendas of Belt and Road. With prominent political scientists such as Zhang Weiwei arguing that China's "super-rich historical and cultural heritages enable it to evolve and develop along its own logic, rather than following the suit of other countries," interpreting the Silk Roads against current debates about *tianxia* will provide a good barometer for assessing how China views itself as a civilizational power, and the structures of moral leadership or hegemony that flow from that.[26]

In their exploration of the historical experience of state rule in Eurasia, Brook, van Praag, and Boltjes argue that the political past is best framed as a series of interpolity relations, and that in the case of China, "every historical regime . . . conducted its relations according to a coherent system that was explicitly hierarchical and centered on that regime as the system's apex and hegemon."[27] The analytical themes pursued here demonstrate how Chinese geocultural power forms part of the strategic displacement of American influence across Asia and beyond today. The question remains, then, will *Pax Americana* be supplanted by new, yet historically familiar, forms of cultural and political hegemony, as Daniel Bell suggests, or will the arguments of Yan Xuetong and other Chinese thinkers carry weight in forging new norms in international relations built around peace and an understanding of equality within hierarchical relationships?[28] The different points of discussion offered here point to the ongoing presence of both. Clearly, resolving the schisms in the analysis of China's growing power is beyond the scope of the discussion here; my more modest aim is to highlight the importance of Belt and Road—as a geocultural, geopolitical phenomena—to these debates.

By framing the "revival" of the Silk Roads in such terms, we can also see how it operates as a "strategic narrative," as identified by Miskimmon, O'Loughlin, and Roselle, one that is designed to influence affairs at both home and abroad. Strategic narratives, they suggest, "are a means for political actors to construct a shared meaning of the past, present, and future of international politics to shape the behavior of domestic and international actors. Strategic narratives are a tool for political actors to extend their influence, manage expectations, and change the discursive environment in which they operate. They are narratives about both states and the system itself, both about who we are and what kind of order we want. The point of strategic narratives is to influence the behavior of others."[29]

A framework of heritage diplomacy captures the asymmetries, both cultural and political, which inhabit the forms of international cooperation created by a Silk Road strategic narrative. Once driven by China, it imposes some clear hierarchies on the worldview constructed through Belt and Road. As we have seen, at one level the Silk Road is merely a story of long-distance trade in silk, but once infused with a host of other values, ideas, and historical events, it becomes a paradigm through which order is constructed, with each actor allocated a place in the system. Multilateral banking structures, regional trade alliances, and bilateral infrastructure agreements solidify that order going forward. But Belt and Road debates in India reveal the ambivalence of many commentators across the region regarding this emergent order and their place therein. For some, such as Ravi Bhoothalingam, the narrative of the historical Silk Road signals a future of Asian co-prosperity built around cooperation: "'I have not told half of what I saw' said Marco Polo, describing the wonders of the Silk Road while dictating his memoirs to his secretary. The ancient roads he traversed led to mythical lands of milk and honey; the Silk Roads of the modern age too could lead to different wonders that are beneficial to all. India and China will have to work together to create many of them, which poses a new challenge to the leadership of both countries. If they can rise up to it, the results could be truly transformative for the world."[30]

But I cite the work of Rieber and Brook, Praag, and Boltjes as a reminder of how these geocultural imaginaries can also be the foundation for new interpolity rivalries.[31] Indeed, there are many in India and elsewhere expressing strong concerns over Beijing's growing political and military power, and the possible ways these might be yielded in the future. With skepticism surrounding China's soft-power strategies, Belt and Road is being scrutinized today for its world-ordering ambitions. In the wake of its 2013 launch, the government of India embarked on its own regional initiatives. The first, Project Mausam, utilized Indian Ocean monsoon histories as a framework for building trade and security alliances across the region. In 2015, a year later, this was followed by the Cotton Route. Similarly oriented around the littoral states of the Indian Ocean, the more expansive Cotton Route was conceived to include the "continental gateways" of Iran and South Africa, linking India to resources in Russia and the African continent, respectively.[32] And in the Palmyra performance of the Mariinsky Theatre orchestra, we see Russia staking its claim as the heir to Eurasia's antiquity and civilization at a time of intense geopolitical competition.[33] Together, then, these various initiatives represent an attempt to build a "transnational collective consciousness" around certain ideas and concepts, one of the core qualities, Hannerz argues, of geocultural imaginaries.[34] But in this convergence of the geocultural and the geopolitical as explored here, we are also seeing some distinct fault lines appear as Asia's "great powers" vie for influence and regional primacy.[35]

The Smooth Touch of Silk

To help understand such processes, my arguments have pivoted around the concept of heritage diplomacy. I see this as a process of smoothing, one that can be contrasted with Anna Tsing's notion of friction as a metaphor for understanding the global connections of today. For Tsing, friction helps reveal "the awkward, unequal, unstable, and creative qualities of interconnection across difference."[36] It adds to the interpretation of globalization as flows by highlighting the grip of encounter, whereby "heterogeneous and unequal encounters can lead to new arrangements of culture and power."[37] The themes pursued here, though, reveal techniques of smoothing in encounters, which produce and rearrange culture and power. In the heritage diplomacy of the Silk Roads, smoothness is aggregated. The topographical smoothing of Richthofen's "original" Silk Road has been greatly amplified in the corridor cartography of Belt and Road. As the overland and maritime routes and their five development corridors have been discursively and visually constructed, extraordinarily formidable terrains and physical environments are subsumed within a grand strategy of economic and physical integration. Belt and Road is a story of hardware and infrastructure, and the challenges of installing and maintaining roads, pipelines, train stations, refineries, and so forth, across deserts and mountains is immense and loaded with unpredictability. The corridors thus conceived work as smoothing devices, creating visions for infrastructure construction across highly uneven and unpredictable physical and political landscapes alike.

Belt and Road crosses politically troubled terrains, and the stress placed on building trust and security is about reducing friction. Such frictions arise in many forms: between partners, but also between projects and extraneous factors, as Tsing highlights. In this respect, the cultural relations Belt and Road activates require careful attention. The people-to-people component is explicitly designed to smooth away frictions with communities affected by the infrastructure projects, and to smooth over the conflicts that arise when lands are claimed, resources are captured and when livelihoods are forever changed. As Belt and Road evolves, culture and cultural heritage will undoubtedly be the source of tensions and a space where violence is inflicted and resisted. As chapter 2 noted, states typically manifest history as heritage by locating the roots of identities in land, borders and the distinctions between mono- or multicultural populations. They are less predisposed to celebrate routes and hybridities. In fact, as we have seen in the case of Xinjiang, routes threaten states. As Millward reminds us: "Xinjiang has been a contact zone for nomad and farmer; for Central Asian, Persian, Turkic and Sinic languages and cultures; for Russian/Soviet and Qing/PRC realms; for Buddhist, Islamic and Communist governing ideologies. At times those contacts have been violent, at times characterised by accommodation."[38]

Xinjiang, Tibet, and Yunnan are thus among the regions in China where the domains of culture and heritage will continue to operate as both friction and smoothing. Looping back to the discussion of China as a civilizational power, it is likely that a "civilizing mission" imperative will be upheld by the cultural dynamics of a Silk Road revival, as heritage becomes a vehicle for imposing the ideologies of progress, development, and the "civilities" of mainstream Han culture on minorities. Belt and Road will also produce parallel situations in other countries across the region.[39] But as we have seen throughout this book, the language of heritage is also being deployed as an ameliorative and facilitator of wider relations. In varying degrees, populations and governments across Asia remain nervous about China's growing influence on the region. The cultural diplomacy of people-to-people exchanges speak to such concerns. But the real strategic value of the historical Silk Roads is that they provide a metaphor for noninterference. Remodeled for the twenty-first century, the Silk Roads are a heritage of trade and exchange that joins proximate and distant places alike in ways that do not challenge the political geographies they connect. Heritage diplomacy also represents one small part of China's efforts to offset any great-power tensions in the region. Asian geopolitics is unlikely to run smoothly any time soon, as Russia, India, Japan, and China all jostle for position and the security that comes from regional influence. As we have seen in chapter 1, the Indian Ocean and East Asia are militarizing rapidly, and Beijing's desire to see Belt and Road succeed necessitates a sustained multidirectional program of diplomacy. The language of trust, harmony, and win-win cooperation is an attempt to reduce tensions at a time when North Korea, Pakistan, Iran, and others are potential sources of instability for the wider Asia region. The cultural components of Belt and Road thus form part of a strategy to advance a model of peace predicated on open borders and multilateral trade. They also form part of a strategy of Chinese internationalism based on an idealized vision of an open China. The themes explored in chapters 3 and 5 indicate how this leads to the reification of the Ming and Tang dynasties. A silken narrative of inter- and intraregion cooperation serves this agenda. If history gives countries like Sri Lanka and Kenya competitive advantage in the networked economy of today, then we should observe how heritage diplomacy is more than just about facilitating relations and attend to its capacity for reconfiguring trade and international relations and the way space is thought about. Commemorating historical connections that predate debates about common laws of the sea and exclusive economic zones by several centuries is an attempt to smooth out tensions over the South China Sea and Indian Ocean.

But as we have seen in the examples of Zheng He, traveling museum exhibitions, and world heritage collaborations, a Silk Roads heritage in an era of Belt and Road involves a second form of smoothing, that of the past itself. Bloodshed in Central Asia, Mongol naval campaigns in Southeast Asia, and episodes of piracy are all glossed over, such that history is pacified and realigned in ways that render it compatible with the international trade and diplomacy of today. The various threads explored in this book have revealed how the material culture of porcelain, shipwrecks, silk, and historical cities and the discourse of diplomacy enact this at a continental scale. Indeed, as we have seen, over a number of decades the Silk Roads have emerged as a quintessential example of Parag Khanna's "mega-diplomacy," one that is now advancing the connectographies of the twenty-first century.[40] It is likely, then, that scholars and cultural-sector institutions will be exposed to, and possibly become part of, the political apparatus of smoothing. Maritime archaeology is thus in for a complex future, at once both a venue for international cooperation and an asset for territorial claims. Moreover, as "partnerships" in Silk Road heritage continue to proliferate, their asymmetries facilitate hegemonic values to be normalized in the international system, such that great powers were always great powers. As the previous chapter noted, the Silk Road gives visibility to much overlooked aspects of Eurasia's past, but it does so in ways that lead to more Sinocentric histories of globalization. It is likely that this will agitate regional neighbors as much as it smooths the path to cooperation.

The Broad Arc of History

What we see here, then, is the convergence of two spatial arcs—the Silk Roads of past centuries and the Belt and Road geographies of the twenty-first century—both of which are constituted by indeterminate networks of connectivity. The future will undoubtedly see additional chapters written in the biographies of the overland and maritime Silk Roads. Both on- and offshore, further evidence of trade, settlement, and cultural exchange awaits discovery. In the story of the Silk Roads there are multiple claims over when they began and finished, where they reached, and who made them flourish. In reactivating these historical representations, Belt and Road expedites the stories and materialities of connectivity in both evidential and creative ways.

This has major implications for how we think about current heritage making practices and policies in the region, in their scale and how they come to be constituted in particular places. The mantra of Belt and Road is routes, channels, hubs, corridors, and opening up—the very underpinnings of neoliberal globalization. But what does this mean culturally? It certainly seems to transform history and heritage in fascinating ways.

As a politics of heritage, Belt and Road represents a moment of historic significance. It was noted in chapters 1 and 2 that states across the wider Asia region have, to date, appropriated their pasts as an instrument of nation building. The arrival of Belt and Road does not profoundly rupture this, but it does signal a new era of bilateral and multilateral heritage making. It is a political economy that changes how history is framed, whereby networked, connected pasts decenter the singular state, and ethno-cultural nationalisms are brought into contact with the cultures of diplomatic relations. Chapter 6 reflected on the implications of this vis-à-vis the potential for new narratives of urban and world history to emerge. Given that history, religion and heritage continue to be a source of conflict and divisiveness in the region, Silk Road narratives undoubtedly represent an important mechanism through which more productive forms of dialogue can be advanced. As we have seen, the involvement of multiple governments gives significant impetus to UNESCO's long-held ambitions of fostering forms of intercultural and international relations.

But to interpret such developments at a more conceptual level Anna Tsing's other recent work, on the precarious economies of the matsutake mushroom is instructive here. In tracing how this Japanese delicacy is bound up in transnational networks involving multiple actors —human and nonhuman—Tsing talks of a "mosaic of open-ended assemblages of entangled ways of life," that in understanding how these come together, we need to look to the various "temporal rhythms and spatial arcs" they create. The Silk Roads, as they are reconstituted as heritagescapes, hold the potential for creating multiple assemblages across Eurasia and parts of Africa. They thus point to the need for a new analytical register, a framework that moves beyond particular sites or national boundaries to capture the ways in which this reimagining of the Silk Roads is emerging through fluid, ambiguous histories, through rolling timelines and via spacious geographies of connectivity. The spatial arcs of the two Silk Roads are indeterminable, and as such infinitely reproducible. Since the launch of Belt and Road, we are seeing the consequences of this historical indeterminacy, as places across the region have begun to proclaim their place within a history of regional connectivity and Silk Road trade. The ongoing recreation of a Silk Roads past thus needs to be understood as more than a set of discursive relations that find their stability in certain places and material forms. As Anna Tsing suggests, "they show us potential histories in the making." From Southeast Asia to Egypt, archaeological material is being dug up or reinterpreted to build evidence of cultural influences from afar and to establish locations as places of passage, as stopping-off points, or as places of commerce and cultural vitality. The Silk Roads narrative encompasses vast territories and incorporates disparate places into a single arc of geography and history. We can thus read how the actual "heritage" of flows, connections, and networks comes to be assembled via a range of human and nonhuman actors—private donors, museum curators, public intellectuals, porcelain, archaeological sites, navigation instruments, or the remnants of ship hulls—all of which gather through and around one another. But as we have seen in the examples of Sri Lanka, Iran, Kenya, and Northwest China, these in turn form part of larger assemblages of regional trade and diplomacy, involving foreign ministers, multinational corporations, airports, container ships, and bilateral trade documents.

But the geopolitical and diplomatic dynamics of Belt and Road add further instability to this picture, as cities and governments creatively use the rhetoric and mechanisms of heritage to embed themselves in the new Silk Roads of the twenty-first century. The material past is thus tweaked and reworked for connectivity, both in the past and in the present. In southern China the exhuming of graves provides the evidence; in Xuanquanzhi in northern China bottom scrapers give credence, and in Lamu pottery sherds offer testimony. In their sheer scale and ambiguity, the Silk Roads tender a narrative to which a seemingly boundless variety and quantity of material evidence can be attached to advance diplomatic interests today. And as the arc of Belt and Road continues to be created, it is likely that the heritage assemblages cited in the preceding chapters will continue to proliferate across the region. The knowledge, practices, and materials of Silk Road heritage will move across locations. And as the political and economic configurations of Belt and Road stabilize, they topographically and thematically redraw how the past is narrated. But here authorship plays a critical role. In chapter 2 we saw Japanese institutions and scholars on either side of World War II weaving Japan and cities such as Nara into the story. As Millward notes, Hillary Clinton's 2011 strategic plan for a north-south Silk Road corridor left out Iran and China, given that US geopolitical affinities lay with South Asia, most notably India.[41] But in the hands of China, the geographies of the Silk Road change again. With Xi'an declared its eastern end, Japan and the Korean Peninsula are excluded from the story, and through a Maritime Silk Road that stretches down to Southeast Asia, China sees itself once again emerging as the Middle Kingdom.

By attending to the logics of Belt and Road, interpretations—indeed, predictions—can be made as to why assemblages of Silk Road heritage form in certain locations and not others. Belt and Road will ensure that maritime heritage industries are cultivated across numerous coastal cities, as far apart as North Africa and Southeast Asia. Clearly, Singapore has certain political, economic, and physical qualities that central Afghanistan does not, and it is important to read how the emergence of a Silk Road heritage industry is contingent on such factors. The argument, then, is that the formation of Silk Road histories across the Eurasia region in the coming years will find their strongest vitality and energy in those locations where they operate as an actor in the larger diplomatic and economic relations of Belt and Road. Smaller-scale assemblages will occur where the political impetus is weaker, nascent, or absent. Belt and Road will not lead to smooth roads of development. It will never be a linear journey of coprosperity. Wars and conflicts will intervene. Instability and uncertainties, as much as grand ambition and coherence, will define the future. What is clear, though, is that we should follow the threads of a politics of silk that carries more than a sheen of historic consequence.

APPENDIX

Key BRI-Related Networks, Associations, Think Tanks, and Research Institutes Established 2014–2016

Networks and Associations

New Silk Road Law Schools Alliance
Silk Road Think Tank Association
Silk Road Think Tank Network (eSiLKS)
Silk Road Universities Network
United Nations Maritime-Continental Silk Road Cities Alliance
University Alliance of the New Silk Road

Research Institutes and Think Tanks

Cambodia 21st Century Maritime Silk Road Research Center, Phnom Penh
Chinese Think Tank Cooperation Alliance for the "Belt and Road," Beijing
City University of Hong Kong Research Centre on One-Belt-One-Road, Hong Kong
Collaborative Innovation Center of Silk Road Economic Belt Research—Xi'an Jiaotong University, Xi'an
Confucius Institute of Maritime Silk Road, Thailand
Council of Cooperative "The Belt and Road Initiative" Think Tank Association
European Think Tank Network on China, Paris

Institute of Maritime Silk Road of Huaqiao University, Quanzhou
International Think Tank for Landlocked Developing Countries, Ulaanbaatar
Macau "One Belt, One Road" Research Center, Macau
Malaysian Institute of Strategic and International Studies (member of SiLKS), Kuala Lumpur
New Silk Road Institute Prague, Prague
One Belt One Road Institute—The Center for China and Globalization, Beijing
One Belt One Road Research Institute—Shanghai Lixin University of Accounting and Finance, Shanghai
Peking University One Belt and One Road Research Centre, Peking
Research Center of Silk Road of Beijing Jiaotong University, Beijing
Silk Road Economic Development Research Centre, Hong Kong
Silk Road Research Center of International Ataturk-Alatoo University, Kyrgyzstan
Silk Road Research Institute—Chinese Academy of Social Sciences, Beijing
Silk Road Research Institute of Beijing Foreign Studies University, Beijing

Notes

WORK TOGETHER FOR A BRIGHT FUTURE

1. This excerpt is from Xi Jinping, "Work Together for a Bright Future of China-Iran Relations," Ministry of Foreign Affairs of the People's Republic of China, January 21, 2016, http://www.fmprc.gov.cn/mfa_eng/wjdt_665385/zyjh_665391/t1334040.shtml.

CHAPTER ONE

1. Embassy of the People's Republic of China in the Islamic Republic of Pakistan, "China-Pakistan Friendship: As Pure and Sincere as the Ever-Flowing Water," November 25, 2015, http://pk.china-embassy.org/eng/zbgx/t1318449.htm.
2. Benedict Anderson, *Imagined Communities: Reflections on the Origin and Spread of Nationalism* (Ithaca, NY: Cornell University Press, 1991).
3. For an overview of this, see Rodney Harrison, *Understanding the Politics of Heritage* (Manchester: Manchester University Press, 2010).
4. Robert Bevan, *The Destruction of Memory: Architecture at War* (London: Reaktion Books, 2006).
5. It is worth pausing here to consider the distinction between *politics* and *the political*. As Barry has indicated, politics involves recognizing and "codifying particular institutional and technical practices" as governments, social movements, or the institutions of parliament. In contrast, "the political need not be only associated with the control of political institutions" and instead refers to "the ways in which artefacts, activities or practices become objects of contestation."

Much of the debate, then, within the field of the politics of heritage has primarily been engaged with the political. See Andrew Barry, *Political Machines: Governing a Technological Society* (London: A & C Black, 2001), 5–6, 201.

6. See, e.g., David Shambaugh, *China Goes Global: The Partial Power* (Oxford: Oxford University Press, 2013); and Joseph Nye, *The Future of Power* (New York: Public Affairs, 2011).
7. Timothy Brook, Michael van Walt van Praag, and Miek Boltjes, eds., *Sacred Mandates: Asian International Relations since Chinggis Khan* (Chicago: University of Chicago Press, 2018).
8. Tim Winter, *Post-Conflict Heritage, Postcolonial Tourism: Culture, Politics and Development at Angkor* (London: Routledge, 2007).
9. Marcel Mauss, *The Gift* (London: Routledge, 2002).
10. For further details, see Natsuko Akagawa, *Heritage Conservation and Japan's Cultural Diplomacy: Heritage, National Identity and National Interest* (London: Routledge, 2014).
11. Akagawa, *Heritage Conservation and Japan's Cultural Diplomacy.*
12. In the case of Myanmar, for example, proposed collaborations around heritage preservation in recent years have led to a revisiting of the idea that the country sits at the "crossroads" of two great civilizations. For further details, see Tim Winter, "Heritage Conservation Futures in an Age of Shifting Global Power," *Journal of Social Archaeology* 14, no. 3 (2014): 319–39.
13. Jeffrey Reeves, "Origins, Intentions, and Security Implications of Xi Jinping's Belt and Road Initiative," in *The Routledge Handbook of Asian Security Studies*, ed. Sumit Ganguly, Andrew Scobell, and Joseph Liow Chin Long (London: Routledge, 2018), loc. 2681 of 14784, Kindle.
14. Reeves, "Origins, Intentions, and Security Implications of Xi Jinping's Belt and Road Initiative," loc. 2688.
15. Reeves, "Origins, Intentions, and Security Implications of Xi Jinping's Belt and Road Initiative," loc. 2688.
16. Reeves, "Origins, Intentions, and Security Implications of Xi Jinping's Belt and Road Initiative," loc. 2695.
17. For further details, see Nadège Rolland, *China's Eurasian Century? Political and Strategic Implications of the Belt and Road Initiative* (Seattle: National Bureau of Asian Research, 2017), 48–52.
18. Wu Jianmin, "'One Belt and One Road,' Far-Reaching Initiative," *China-US Focus*, March 26, 2015, http://www.chinausfocus.com/finance-economy/one-belt-and-one-road-far-reaching-initiative/.
19. "Vision and Actions on Jointly Building Silk Road Economic Belt and 21st-Century Maritime Silk Road, Issued by the National Development and Reform Commission, Ministry of Foreign Affairs, and Ministry of Commerce of the People's Republic of China, with State Council authorization," National Development and Reform Commission—People's Republic of

China, last modified March 28, 2015, http://en.ndrc.gov.cn/newsrelease/201503/t20150330_669367.html.

20. See "China Invests $124bn in Belt and Road Global Trade Project," *BBC News*, May 14, 2017, http://www.bbc.com/news/world-asia-39912671.
21. He Yini, "China to Invest $900b in Belt and Road Initiative," *China Daily USA*, May 28, 2015, http://usa.chinadaily.com.cn/business/2015-05/28/content_20845687.htm.
22. Michele Ruta and Mauro Boffa, "Trade Linkages among Belt and Road Economies: Three Facts and One Prediction," *The Trade Post* (blog), May 31, 2018, https://blogs.worldbank.org/trade/trade-linkages-among-belt-and-road-economies-three-facts-and-one-prediction.
23. For further details, see Michael Clarke, "The Belt and Road Initiative: China's New Grand Strategy?," *Asia Policy* 24 (2017): 71–79.
24. Rachel Brown, "Where Will the New Silk Road Lead? The Effects of Chinese Investment and Migration in Xinjiang and Central Asia," *Columbia University Journal of Politics and Society* 26 (2016): 72–74.
25. UNESCO is the UN Educational, Scientific, and Cultural Organization. See UNESCO Kazakhstan, "At UNESCO, Kazakhstan's President Nazarbayev Calls for Intercultural Dialogue to Counter Extremism," http://www.unesco.kz/new/en/unesco/news/2998.
26. Isabelle Meere, "Asian Leaders Clash over Belt and Road at SCO Summit," Centre of Expertise on Asia, June 12, 2018, https://asiahouse.org/asian-leaders-clash-belt-road-sco-summit/.
27. See, e.g., Ananth Krishnan, "China Offers to Develop Chittagong Port," *The Hindu*, March 15, 2010, http://www.thehindu.com/news/international/china-offers-to-develop-chittagong-port/article245961.ece; and Shannon Tiezzi, "Chinese Company Wins Contract for Deep Sea Port in Myanmar," *The Diplomat*, January 1, 2016, http://thediplomat.com/2016/01/chinese-company-wins-contract-for-deep-sea-port-in-myanmar/.
28. Kent Calder, *The New Continentalism: Energy and Twenty-First-Century Eurasian Geopolitics* (New Haven, CT: Yale University Press, 2012); and Chris Devonshire-Ellis, *China's New Economic Silk Road: The Great Eurasian Game and the String of Pearls* (Hong Kong: Asia Briefing, Dezan Shira and Associates, 2015).
29. For further details, see "BD, China Sign Deal to Build SEZ in Ctg," *Financial Express*, May 3, 2017, http://www.thefinancialexpress-bd.com/2016/06/17/34490/BD,-China-sign-deal-to-build-SEZ-in-Ctg; and Viola Zhou, "China Wins US$3 Billion Bid to Build Rail Line in Bangladesh," *South China Morning Post*, August 9, 2016, http://www.scmp.com/business/companies/article/2001471/china-wins-us3-billion-bid-build-rail-line-bangladesh.
30. Amy Chew, "China, Malaysia Tout New 'Port Alliance' to Reduce Customs Bottlenecks and Boost Trade," *South China Morning Post*, April 9, 2016,

http://www.scmp.com/news/asia/southeast-asia/article/1934839/china-malaysia-tout-new-port-alliance-reduce-customs; and Robin Bromby, "Down the Maritime Silk Road," *The Australian*, December 6, 2013, http://www.theaustralian.com.au/business/in-depth/down-the-maritime-silk-road/story-fnjy4qn5-1226776242929.

31. In late 2016, the development-aid tracking project AidData reported more than two hundred active Chinese assistance projects in Cambodia. See AidData's "Geospatial Dashboard," accessed October 2, 2016, http://china.aiddata.org. See also "Cambodia Positions Itself along New Silk Road," *Cambodia Daily*, June 27, 2016, https://www.cambodiadaily.com/news/cambodia-positions-itself-along-new-silk-road-114629/; and Ben Paviour, "For China, 'Cambodia Is a Sideshow, but It's a Loyal One,'" *Cambodia Daily*, October 19, 2016, https://www.cambodiadaily.com/news/china-cambodia-sideshow-loyal-one-119475/.
32. See Sudha Ramachandran, "Iran, China and the Silk Road Train," *The Diplomat*, March 20, 2016, http://thediplomat.com/2016/03/iran-china-and-the-silk-road-train/.
33. "Chabahar Port to Harbor Chinese Industrial Town," Press TV, April 27, 2016, http://www.presstv.com/Detail/2016/04/27/462797/China-CMI-Iran-mega-port.
34. See, e.g., "Iran, China Agree $600 Billion Trade Deal after Sanctions," *Dawn*, January 23, 2016, http://www.dawn.com/news/1234923.
35. See, e.g., Saeed Shah, "China to Build Pipeline from Iran to Pakistan," *Wall Street Journal*, April 9, 2015, http://www.wsj.com/articles/china-to-build-pipeline-from-iran-to-pakistan-1428515277.
36. David Arase, "China's Two Silk Roads Initiative: What It Means for Southeast Asia," *Southeast Asian Affairs* 2015, no. 1 (2015): 28.
37. Russia's initial reluctance to be part of BRI, and its subsequent embrace of the project, is detailed by Sebastien Peyrouse, "The Evolution of Russia's Views on the Belt and Road Initiative," *Asia Policy* 24 (2017): 96–102.
38. For further details, see Stephen Blank, "The Dynamics of Russo-Chinese Relations," in *The Routledge Handbook of Asian Security Studies*, ed. Sumit Ganguly, Andrew Scobell, and Joseph Liow Chin Long (London: Routledge, 2018), Kindle.
39. See, e.g., Clarke, "Belt and Road Initiative," 71–79.
40. Arase, "China's Two Silk Roads Initiative," 33.
41. Anthony Milner, "Culture and the International Relations of Asia," *Pacific Review* 30, no. 6 (2017): 857–69.
42. Graham Allison, *Destined for War: Can America and China Escape Thucydides's Trap?* (London: Scribe, 2017), Kindle.
43. They state: "While much research has addressed China's rising power and the global implications thereof, there is no analytical framework for illustrating and interpreting the four coexisting phenomena that characterize China's global power strategy: China's self-identification as a

great power; the conscious linking of trade and diplomacy; intertwined economic openness, higher education and cultural assimilation; and China's sensitivity to international criticism." Su-Yan Pan and Joe Tin-yau Lo, "Re-conceptualizing China's Rise as a Global Power: A Neo-Tributary Perspective," *Pacific Review* 30, no. 1 (2017): 6.

44. Pan and Lo, "Re-conceptualizing China's Rise as a Global Power."
45. See, e.g., Andrea Forsby, "An End to Harmony? The Rise of a Sino-Centric China," *Political Perspectives* 5, no. 3 (2011): 5–26; and David Shambaugh, *China Goes Global: The Partial Power* (Oxford: Oxford University Press, 2013).
46. See, e.g., "After 'Chinese Dream,' Xi Jinping Outlines Vision for 'Asia-Pacific Dream' at APEC Meet," *South China Morning Post*, November 10, 2014, http://www.scmp.com/news/china/article/1635715/after-chinese-dream-xi-jinping-offers-china-driven-asia-pacific-dream.
47. In 2011 Xi Jinping dedicated a plenary session of the Seventeenth Central Committee of the Chinese Communist Party to the issue, and he has stated that the China Dream involves the renaissance of Chinese civilization. For further details, see Shambaugh, *China Goes Global*.
48. See, e.g., Yan Xuetong, "From Keeping a Low Profile to Striving for Achievement," *Chinese Journal of International Politics* 7, no. 2 (2014): 153–84.
49. Jeff Adams, "The Role of Underwater Archaeology in Framing and Facilitating the Chinese National Strategic Agenda," in *Cultural Heritage Politics in China*, ed. Tami Blumenfield and Helaine Silverman (New York: Springer, 2013). See also Edward Vickers, "A Civilizing Mission with Chinese Characteristics? Education, Colonialism and Chinese State Formation in Comparative Perspective," in *Constructing Modern Asian Citizenship*, ed. Edward Vickers and Krishna Kumar (Oxford: Routledge, 2015), 50–79.
50. Andrew Scobell, "Whither China's 21st Century Trajectory?," in *The Routledge Handbook of Asian Security Studies*, ed. Sumit Ganguly, Andrew Scobell, and Joseph Liow Chin Long (London: Routledge, 2018), loc. 948 of 14784, Kindle.
51. See "Asia's Era of Infrastructure," video from World Economic Forum annual meeting, January 22, 2016, https://www.weforum.org/events/world-economic-forum-annual-meeting-2016/sessions/asia-s-era-of-infrastructure/.
52. UNESCAP is the UN Economic and Social Commission for Asia and the Pacific.
53. TRACECA is Transport Corridor Europe-Caucasus Asia.
54. Rolland, *China's Eurasian Century?*, 9.
55. For a concise and helpful overview of the unilateral and multilateral regional initiatives that precede BRI, see Rolland, *China's Eurasian Century?*.
56. Kearrin Sims, "The Asian Development Bank and the Production of Poverty: Neoliberalism, Technocratic Modernization and Land Dispossession

in the Greater Mekong Subregion," *Singapore Journal of Tropical Geography* 36, no. 1 (2015): 112–26.

57. Rolland, *China's Eurasian Century?*, 41.
58. Emma Mawdsley, "Development Geography 1: Cooperation, Competition and Convergence between 'North' and 'South,'" *Progress in Human Geography* 41, no. 1 (2017): 108–17.
59. Smruti Pattanaik, "Indian Ocean in the Emerging Geo-Strategic Context: Examining India's Relations with Its Maritime South Asian Neighbours," *Journal of the Indian Ocean Region* 12, no. 2 (2016): 126–42.
60. For further details, see Sangeeta Khorana and Leila Choukroune, "India and the Indian Ocean Region," *Journal of the Indian Ocean Region* 12, no. 2 (2016): 122–25.
61. For further details, see Jivanta Schöttli, "Editorial—Special Issue: Power, Politics and Maritime Governance in the Indian Ocean," *Journal of the Indian Ocean Region* 9, no. 1 (2013): 1–5.
62. Antara Singh, "India, China and the US: Strategic Convergence in the Indo-Pacific," *Journal of the Indian Ocean Region* 12, no. 2 (2016): 8.
63. Christian Bouchard and William Crumplin, "Neglected No Longer: The Indian Ocean at the Forefront of World Geopolitics and Global Geostrategy," *Journal of the Indian Ocean* 6, no. 1 (2010): 26.
64. David Brewster, "An Indian Ocean Dilemma: Sino-Indian Rivalry and China's Strategic Vulnerability in the Indian Ocean," *Journal of the Indian Ocean Region* 11, no. 1 (2015): 57.
65. David Brewster, "Beyond the 'String of Pearls': Is There Really a Sino-Indian Security Dilemma in the Indian Ocean?," *Journal of the Indian Ocean Region* 10, no. 2 (2014): 133–49; and Brewster, "Indian Ocean Dilemma," 48–59.
66. Brewster, "Beyond the 'String of Pearls.'"
67. See Brewster, "Indian Ocean Dilemma," 48–59, for further examples of overland infrastructure projects that find points of connection with the Indian Ocean.
68. See, e.g., Schöttli, "Editorial"; and Isabel Hofmeyer, "Styling Multilateralism: Indian Ocean Cultural Futures," *Journal of the Indian Ocean Region* 11, no. 1 (2015): 98–109.
69. For further details, see War Memoryscapes in Asia Project (WARMAP), "What Is Warmap?," http://www.warinasia.com.
70. Tim Winter, "Heritage Diplomacy," *International Journal of Heritage Studies* 21, no. 10 (2015): 997–1015; Amy Clarke, "Heritage Diplomacy," in *Handbook of Cultural Security*, ed. Yasushi Watanabe (Cheltenham, UK: Edward Elgar Publishing, 2018), 417–36; Akagawa, *Heritage Conservation and Japan's Cultural Diplomacy*; and Christina Luke and Morag Kersel, *U.S. Cultural Diplomacy and Archaeology: Soft Power, Hard Heritage* (New York: Routledge, 2013).
71. Thomas Weiss, *Global Governance: Why? What? Whither?* (Cambridge, UK: Polity, 2013), 143; Lynn Meskell, Claudia Liuzza, Enrico Bertacchini,

and Donatella Saccone, "Multilateralism and UNESCO World Heritage: Decision-Making, States Parties and Political Processes," *International Journal of Heritage Studies* 21, no. 5 (2015): 423–40; Lynn Meskell, "States of Conservation: Protection, Politics, and Pacting within UNESCO's World Heritage Committee," *Anthropological Quarterly* 87, no. 1 (2014): 217–43; and Lynn Meskell, "The Rush to Inscribe: Reflections on the 35th Session of the World Heritage Committee UNESCO Paris, 2011," *Journal of Field Archaeology* 37, no. 2 (2012): 145–51.

72. Andrew Cooper, Jorge Heine, and Ramesh Chandra Thakur, eds., *The Oxford Handbook of Modern Diplomacy* (Oxford: Oxford University Press, 2013).

73. Stuart Murray, Paul Sharp, Geoffrey Wiseman, David Criekemans, and Jan Melissen, "The Present and Future of Diplomacy and Diplomatic Studies," *International Studies Review* 13, no. 4 (2011): 711.

74. Alister Miskimmon, Ben O'Loughlin, and Laura Roselle, *Strategic Narratives: Communication Power and the New World Order* (London: Routledge, 2013), Kindle; and Edward Wastnidge, "Strategic Narratives and Iranian Foreign Policy into the Rouhani Era," *e-International Relations*, March 10, 2016, https://www.e-ir.info/2016/03/10/strategic-narratives-and-iranian-foreign-policy-into-the-rouhani-era/.

75. Manuel Castells and Gustavo Cardoso, eds., *The Network Society: From Knowledge to Policy* (Washington, DC: John Hopkins Center for Transatlantic Relations, 2005); and Saskia Sassen, *Global Networks, Linked Cities* (London: Routledge, 2002).

76. For assemblage theory, see Gilles Deleuze and Félix Guattari, *A Thousand Plateaus: Capitalism and Schizophrenia* (London: Bloomsbury Academic, 2013); Bruno Latour, *Reassembling the Social: An Introduction to Actor-Network-Theory* (Oxford: Oxford University Press, 2007); John Law, *After Method: Mess in Social Science Research* (London: Routledge, 2004); and Theodore Schatzki, "Materiality and Social Life," *Nature and Culture* 5, no. 2 (2010): 123–49.

77. Ian Hodder, *Entangled: An Archaeology of the Relationships between Humans and Things* (Malden, MA: Wiley-Blackwell, 2012).

78. Andrew Barry, *Material Politics: Disputes along the Pipeline* (Oxford: Wiley-Blackwell, 2013); and Agneishka Joniak-Luthi, "Roads in China's Borderlands: Interfaces of Spatial Representations, Perceptions, Practices, and Knowledges," *Modern Asian Studies* 50, no. 1 (2016): 118–40.

79. Deleuze and Guattari, *Thousand Plateaus*.

80. See, e.g., *Silk Road Countries* (map) (Budapest: GiziMap, 2015).

81. For a more detailed discussion of this issue, see Marie Thorsten, "Silk Road Nostalgia and Imagined Global Community," *Comparative American Studies: An International Journal* 3, no. 3 (2005): 301–17.

82. James Millward, *The Silk Road: A Very Short Introduction* (Oxford: Oxford University Press, 2013), loc. 481 of 2963, Kindle. See also Armin

Selbitschka, "The Early Silk Road(s)," in *Oxford Research Encyclopedia of Asian History,* ed. David Ludden (New York: Oxford University Press, 2018), https://doi.org/10.1093/acrefore/9780190277727.013.2.

83. See, e.g., Robert Collins, *East to Cathay: The Silk Road* (New York: McGraw-Hill Book Co., 1986); Xinru Liu, *The Silk Road in World History* (Oxford: Oxford University Press, 2010), Kindle; and Luce Boulnois, *The Silk Road* (London: George Allen and Unwin, 1966).
84. Liu, *Silk Road in World History,* loc. 1313 of 3296.
85. Liu, *Silk Road in World History,* loc. 1294.
86. As Arabs moved out of tents into more permanent settlements, the tapestries and rugs used to adorn tents were ideal for softening stone and marble surfaces. It was during this period that the tradition of covering the Kaaba in Mecca with black cloth also began, with annual donations required to ensure that its appearance remained pristine. A donation of expensive, carefully decorated silk trims bestowed legitimacy and prestige upon the donor.
87. Boulnois, *Silk Road.*
88. See, e.g., Gary Nathan, *Cumin, Camels and Caravans: A Spice Odyssey* (Berkeley: University of California Press, 2014); and Hermann Parzinger, "The 'Silk Roads' Concept Reconsidered: About Transfers, Transportation and Transcontinental Interactions in Prehistory," *Silk Road* 5, no. 2 (2008): 7–15.
89. Susan Whitfield, *Life along the Silk Road* (Berkeley: University of California Press, 2015), Kindle; and Eric Herbert Warmington, *The Commerce between the Roman Empire and India* (Delhi: Vikas Publishing House, 1928).
90. See Sven Beckert, *Empire of Cotton: A New History of Global Capitalism* (London: Penguin, 2015); John Griffiths, *Tea: A History of the Drink That Changed the World* (London: Andre Deutsch, 2011); and Mark Kurlansky, *Paper: Paging through history* (New York: W. W. Norton and Co., 2017).
91. See, e.g., Thomas Barfield, "Steppe Empires, China, and the Silk Route: Nomads as a Force in International Trade and Politics," in *Nomads in the Sedentary World,* ed. Anatoly Khazanov and Andre Wink (Surrey, UK: Curzon Press, 2001), 234–49; Michal Biran, "Introduction: Nomadic Culture," in *Nomads as Agents of Cultural Change: The Mongols and their Eurasian Predecessors,* ed. Leuven Amitai and Michal Biran (Honolulu: University of Hawai'i Press, 2015), 1–9; Anatoly Khazanov, "Nomads in the History of the Sedentary World," in *Nomads in the Sedentary World,* ed. Anatoly Khazanov and Andre Wink (Surrey, UK: Curzon Press, 2001), 1–23; Selbitschka, "The Early Silk Road(s)"; and Tansen Sen, "Diplomacy, Trade and the Quest for the Buddha's Tooth: The Yongle Emperor and Ming China's South Asian Frontier," in *Ming China: Courts and Contacts, 1400–1450,* ed. Craig Clunas, Jessica Harrison-Hall, and Yu-ping Luk (London: British Museum Press, 2016), 26–36.
92. Millward, *The Silk Road: A Very Short Introduction*; Susan Whitfield, "Was There a Silk Road?," *Asian Medicine* 3 (2007): 201–13; Khodadad

Rezakhani, "The Road That Never Was: The Silk Road and Trans-Eurasian Exchange," *Comparative Studies of South Asia, Africa and the Middle East* 30, no. 3 (2010): 420–33; and Biran, "Nomadic Culture," 1–9.

93. See Richard Foltz, *Religions of the Silk Road: Premodern Patterns of Globalization* (New York: Springer Publishing, 2010).
94. See, e.g., Parzinger, "'Silk Roads' Concept Reconsidered," 7–15.
95. For further details, see Elena Kuzmina, *The Prehistory of the Silk Road*, ed. Victor H. Mair (Philadelphia: University of Pennsylvania Press, 2008), 8–17.
96. Valerie Hansen, *The Silk Road: A New History with Documents* (Oxford: Oxford University Press, 2017).
97. An excellent example of this approach is offered by Whitfield, *Life along the Silk Road*.
98. During his 1907 visit to Dunhuang, Stein found a copy of the Diamond Sutra, a Mahayana sutra, dating back to the Tang dynasty. See Joyce Morgan and Conrad Walters, *Journeys on the Silk Road* (Guilford, CT: Lyons Press, 2012).
99. In *Night Train to Turkistan*, for example, Stuart Stevens wrote of a trip through China conceived around the journey made by Peter Fleming from Beijing to Kashgar in 1934. For further details, see Stuart Stevens, *Night Train to Turkistan: Adventures along China's Silk Road* (London: Paladin Grafton Books, 1990).
100. See Christopher Beckwith, *Empires of the Silk Road: A History of Central Eurasia from the Bronze Age to the Present* (Princeton, NJ: Princeton University Press, 2009); and John Miksic, *Singapore and the Silk Road of the Sea, 1300–1800* (Singapore: National University of Singapore Press, 2013).
101. Miksic, *Singapore and the Silk Road of the Sea*.

CHAPTER TWO

1. This can be broadly confirmed through a Google Ngram search for the term, which shows a distinct jump in the early 1980s, with further acceleration occurring from the 1990s onward.
2. Kuzmina, for example, argues that the origins of the Silk Road are in the nomadic worlds of the Eurasian Steppe, stretching back three millennia. See Elena Kuzmina, *The Prehistory of the Silk Road*, ed. Victor H. Mair (Philadelphia: University of Pennsylvania, 2008), 8–17.
3. Voll even identifies five "stages" of the Silk Road, with the final one beginning in the nineteenth century with the opening of the Suez Canal. See John Obert Voll, "Main Street of Eurasia," Silk Roads: Dialogue, Diversity, and Development, https://en.unesco.org/silkroad/content/main-street-eurasia.
4. See, e.g., Armin Selbitschka, "The Early Silk Road(s)," in *Oxford Research Encyclopedia of Asian History*, ed. David Ludden (New York: Oxford University Press, 2018), https://doi.org/10.1093/acrefore/9780190277727.013.2.

5. See Edward Ingram, "Great Britain's Great Game: An Introduction," *International History Review* 2, no. 2 (1980): 162.
6. Ingram, "Great Britain's Great Game," 166.
7. Suzanne Marchand, *German Orientalism in the Age of Empire: Religion, Race, and Scholarship* (Cambridge: Cambridge University Press, 2009), 154.
8. Daniel Waugh, "The Making of Chinese Central Asia," *Central Asia Survey* 26, no. 2 (2007): 8.
9. Tamara Chin, "The Invention of the Silk Road, 1877," *Critical Inquiry* 40, no. 1 (2013): 194–219.
10. Chin, "Invention of the Silk Road, 1877."
11. Sven Hedin's expeditions to the region began in the early 1890s with mapping surveys conducted in and around the Taklamakan Desert. He would continue to travel to the region for thirty-five years.
12. For further details on the construction of the Trans-Caspian, see W. E. Wheeler, "The Control of Land Routes: Russian Railways in Central Asia," *Journal of the Royal Central Asian Society* 21, no. 4 (1934): 585–608.
13. The two cities were finally connected in 1916. With the quality of construction suffering from a wholly inadequate budget, upgrades to the line continued throughout the twentieth century.
14. Wheeler, "Control of Land Routes," 605.
15. This expression has been used by numerous authors over the course of the twentieth century. See, e.g., Wheeler, "Control of Land Routes."
16. The article was published in *Geographical Journal* in April the same year. See Halford Mackinder, "The Geographical Pivot of History," *Geographical Journal* 23, no. 4 (1904): 421–37.
17. Sarah O'Hara, "Great Game or Grubby Game? The Struggle for Control of the Caspian," *Geopolitics* 9, no. 1 (2004): 138–60.
18. A detailed description of Kaye's deployment of the term and its subsequent usage by others is offered by Seymour Becker, "The 'Great Game': The History of an Evocative Phrase," *Asian Affairs* 43, no. 1 (2012): 61–80.
19. Ingram and Becker are among those who indicate that the term primarily relates to the British experience of the rivalry in Central Asia. For further details, see Ingram, "Great Britain's Great Game"; and Becker, "'Great Game.'"
20. See Selçuk Esenbel, introduction to *Japan on the Silk Road: Encounters and Perspectives of Politics and Culture in Eurasia*, ed. Selçuk Esenbel (Leiden: Brill, 2018), 1–34.
21. Ian Nish, "Japan and the Great Game," in *Japan on the Silk Road: Encounters and Perspectives of Politics and Culture in Eurasia*, ed. Selçuk Esenbel (Leiden: Brill, 2018), 35–47.
22. Nile Green, "Introduction: Writing, Travel, and the Global History of Central Asia," in *Writing Travel in Central Asian History*, ed. Nile Green (Bloomington: Indiana University Press, 2014), p. 3, Kindle.

23. Russell-Smith has argued this expedition informed Aurel Stein's decision to set off in search of the caves nearly three decades later. See Lilla Russell-Smith, "Hungarian Explorers in Dunhuang," *Journal of the Royal Asiatic Society* 10, no. 3 (2000): 341.
24. Peter Hopkirk, *Foreign Devils on the Silk Road: The Search for the Lost Treasures of Central Asia* (London: John Murray, 1980), loc. 608 of 3743, Kindle.
25. Hopkirk, *Foreign Devils on the Silk Road,* locs. 837–41.
26. Hopkirk, *Foreign Devils on the Silk Road,* loc. 986.
27. Green, "Introduction," 24.
28. Karl Meyer and Sharon Brysac, *Tournament of Shadows: The Great Game and the Race for Empire in Central Asia* (New York: Basic Books, 1999), 360–3.
29. Meyer and Brysac, *Tournament of Shadows,* 111–36.
30. Ingo Strauch, "Priority and Exclusiveness: Russians and Germans at the Northern Silk Road (Materials from the Turfan-Akten)," *Études de lettres* 2–3 (2014): 147–50.
31. Valerie Hansen, *The Silk Road: A New History with Documents* (Oxford: Oxford University Press, 2017); and Frances Wood, *The Silk Road: Two Thousand Years in the Heart of Asia* (Berkeley: University of California Press, 2002).
32. Esenbel, "Introduction," 7–8.
33. Meyer and Brysac, *Tournament of Shadows,* 371.
34. Ōtani later served as the twenty-second patriarch of the Honpa Honganji, a branch of the Jōdo Shinshū sect, and as chief abbot of its head temple, the Nishi Honganji in Kyoto. Küçükyalçin argues that through his leadership of the Western Honganji, Ōtani "played a crucial role in the making of modern Japan." For more detail, see Erdal Küçükyalçin, "Ōtani Kozui and His Vision of Asia: From Villa Nirakusō to 'The Rise of Asia' Project," in *Japan on the Silk Road: Encounters and Perspectives of Politics and Culture in Eurasia,* ed. Selçuk Esenbel (Leiden: Brill, 2018), 181–98 (quote at 181); and Brij Tankha, "Exploring Asia, Reforming Japan: Ōtani and Itō Chūta," in *Japan on the Silk Road,* 156–80.
35. See Imre Galambos, "Japanese 'Spies' along the Silk Road: British Suspicions Regarding the Second Ōtani Expedition (1908–09)," *Japanese Religions* 35, nos. 1–2 (2010): 33–61; Imre Galambos, "Japanese Exploration of Central Asia: The Ōtani Expeditions and Their British Connections," *Bulletin of SOAS* 75, no. 1 (2012): 113–34; and Imre Galambos, "Buddhist Relics from the Western Regions: Japanese Archaeological Exploration of Central Asia," in *Writing Travel in Central Asian History,* ed. Nile Green (Bloomington: Indiana University Press, 2014), pp. 152–69, Kindle.
36. For a more detailed discussion of this trip and the construction of his villa in Kobe, see Tankha, "Exploring Asia, Reforming Japan," 168.
37. Tankha, "Exploring Asia, Reforming Japan."
38. Research for Peter Hopkirk's work was undertaken in the archives of London at Kew and the political and secret files kept in the India Office Library.

39. Galambos, "Japanese 'Spies' along the Silk Road," 33–61.
40. Galambos, "Buddhist Relics from the Western Regions," 152–69.
41. Galambos, "Buddhist Relics from the Western Regions," 166.
42. Tansen Sen, *Buddhism, Diplomacy and Trade: The Realignment of India-China Relations, 600–1400* (Lanham, MD: Rowman & Littlefield, 2016), loc. 293, Kindle.
43. Mark Ravinder Frost, "Handing Back History: Britain's Imperial Heritage State in Colonial Sri Lanka and South Asia, 1870–1920" (keynote address, National Symposium of Historical Studies, University of Sri Lanka, January 31, 2018).
44. For a detailed account of the intersection between pan-Asianism and pan-Islamism, see Cemil Aydin, *The Politics of Anti-Westernism in Asia: Visions of World Order in Pan-Islamic and Pan-Asian Thought* (New York: Columbia University Press, 2007), Kindle.
45. Aydin, *Politics of Anti-Westernism in Asia.*
46. Aydin, *Politics of Anti-Westernism in Asia.*
47. Ali Merthan Dündar, "The Effects of the Russo-Japanese War on Turkic Nations: Japan and Japanese in Folk Songs, Elegies and Poems," in *Japan on the Silk Road: Encounters and Perspectives of Politics and Culture in Eurasia,* ed. Selçuk Esenbel (Leiden: Brill, 2018), 199–227.
48. Esenbel, "Introduction," 33.
49. See, e.g., Penny Edwards, *Cambodge: The Cultivation of a Nation, 1860–1945* (Honolulu: University of Hawai'i Press, 2007); and Tapati Guha-Thakurta, *Monuments, Objects, Histories: Institutions of Art in Colonial and Postcolonial India* (New York: Columbia University Press, 2004).
50. Judith Snodgrass, *Presenting Japanese Buddhism to the West: Orientalism, Occidentalism, and the Columbian exposition* (Chapel Hill: University of North Carolina Press, 2003).
51. See introduction to Kwa Chong-Guan, ed., *Early Southeast Asia Viewed from India: An Anthology of Articles from the "Journal of the Greater India Society"* (Delhi: Manohar, 2013).
52. It is important to note, however, that, much like the vestiges of exchange and encounter found in Central Asia, histories of seafaring and long-distance connections remained marginal in the cultural politics of nation building in Asia for much of the twentieth century.
53. Leo Klejn, *Soviet Archaeology: Schools, Trends and History* (Oxford: Oxford University Press, 2012), 3–12.
54. V. Bulkin, Leo Klejn, and G. S. Lebedev, "Attainments and Problems of Soviet Archaeology," *World Archaeology* 13, no. 3 (1982): 272–95.
55. Klejn, *Soviet Archaeology,* 13–49, 135–42.
56. Pavel Dolukhanov, "Archaeology and Nationalism in Totalitarian and Post-Totalitarian Russia," in *Nationalism and Archaeology: Scottish Archaeological Forum,* ed. John Atkinson, Iain Banks, and Jerry O'Sullivan (Glasgow: Cruithne Press, 1996), 200–213.

57. Bulkin, Klejn, and Lebedev, "Attainments and Problems of Soviet Archaeology," 276.
58. The content of the Soviet nationalities policy was formulated for the Twelfth Party Congress in April 1923. With a resolution passed, the Central Committee added further detail to the policy a few months later. See Terry Martin, *The Affirmative Action Empire: Nations and Nationalism in the Soviet Union, 1923–1939* (Ithaca, NY: Cornell University Press, 2017), Kindle.
59. Martin, *Affirmative Action Empire*, loc. 470 of 820.
60. Francine Hirsch, *Empire of Nations: Ethnographic knowledge and the Making of the Soviet Union* (Ithaca, NY: Cornell University Press, 2014), Kindle.
61. Oksana Sarkisova, *Screening Soviet Nationalities: Kulturfilms from the Far North to Central Asia* (London: I. B. Tauris, 2017), loc. 502 of 6343, Kindle.
62. Sarkisova, *Screening Soviet Nationalities*, loc. 3356.
63. Sarkisova, *Screening Soviet* Nationalities, loc. 3501.
64. Justin Jacobs, "Cultural Thieves or Political Liabilities? How Chinese Officials Viewed Foreign Archaeologists in Xinjiang, 1839–1914," *Silk Road* 10 (2012): 117.
65. Justin Jacobs, "Nationalist China's 'Great Game': Leveraging Foreign Explorers in Xinjiang, 1927–1935," *Journal of Asian Studies* 73, no. 1 (2014): 51.
66. This trip and its background have also featured in other accounts of this period's archaeology. See, e.g., Hopkirk, *Foreign Devils on the Silk Road*; and Helen Wang, "Sir Aurel Stein," in *From Persepolis to the Punjab: Exploring Ancient Iran, Afghanistan and Pakistan*, ed. Elizabeth Errington and Vesta Curtis (London: British Museum Press, 2007), 227–34.
67. For a detailed account of this process, see James Leibold, *Reconfiguring Chinese Nationalism: How the Qing Frontier and Its Indigenes Became Chinese* (New York: Palgrave Macmillan, 2007), Kindle.
68. Justin Jacobs, "Confronting Indiana Jones: Chinese Nationalism, Historical Imperialism and the Criminalisation of Aurel Stein and the Raiders of Dunhuang, 1899–1944," in *China on the Margins*, ed. Sherman Cochran and Paul Pickowicz (Ithaca, NY: Cornell University Press, 2010), 65–90.
69. Jacobs, "Confronting Indiana Jones," 82.
70. Hedin also used the term for a set of recommendations to the Chinese government. His *Plan for the Revival of the Silk Road* included proposals for new air and rail transport corridors linking Europe to Asia. The report was drafted to help advance the interests of Lufthansa and the German government, among others, as the following excerpt reveals: "In my memorandum to the Nanking Government I stressed the magnificence of a revival of the ancient Imperial Highway, the road along which the silk was carried for centuries in an unbroken stream towards the western lands. It was, indeed, to study this link between China proper and the heart of Asia, and find out what it needed in improvement and upkeep to make it usable for motor traffic on a large scale, that we were now in the field. . . . I

myself had no objection to the slowness with which we crawled along the Silk Road. I had plenty of time to observe both the road and the surrounding landscape, town and village life, people and traffic—in a word, reality as it passed before our eyes. But, I will readily confess, I lived most in the world of imagination, in the past with its impressive pictures and seething life, and in the future with its splendid prospects of technical progress and the development of human energy on a scale that makes the brain reel. I have been glad to hear from different sources in China that the Government has already begun this gigantic undertaking." Sven Hedin, *The Silk Road* (London: George Routledge and Sons, 1938): 229–31. See also Chin, "Invention of the Silk Road, 1877," 26–27.

71. Daniel Waugh, "The Silk Roads in History," *Expedition* 52, no. 3 (2010): 13–14.
72. Tim Winter, *The Silk Road: A Biography* (forthcoming).
73. Daniel Waugh, "Richthofen's 'Silk Roads': Towards the Archaeology of a Concept," *Silk Road* 5, no. 1 (2007): 17.
74. See Bruce Trigger, *A History of Archaeological Thought* (Cambridge: Cambridge University Press, 1989); Elizabeth Errington and Vesta Curtis, "The British and Archaeology in Nineteenth-Century Persia," in *From Persepolis to the Punjab: Exploring Ancient Iran, Afghanistan and Pakistan*, ed. Elizabeth Errington and Vesta Curtis (London: British Museum Press, 2007), 166–78; and Elizabeth Errington and Vesta Curtis, "The Explorers and Collectors," in *From Persepolis to the Punjab: Exploring Ancient Iran, Afghanistan and Pakistan*, ed. Elizabeth Errington and Vesta Curtis (London: British Museum Press, 2007), 3–16.
75. Astrid Swenson and Peter Mandler, eds., *From Plunder to Preservation: Britain and the Heritage of the Empire, c. 1800–1940* (Oxford: Oxford University Press, 2013).
76. Benjamin Porter, "Near Eastern Archaeology: Imperial Pasts, Postcolonial Presents, and the Possibilities of a Decolonized Future," in *Handbook of Postcolonial Archaeology*, ed. Jane Lydon and Uzma Z. Rizvi (Oxford: Routledge, 2016), 53.
77. Eric Cline, *Biblical Archaeology: A Very Short Introduction* (Oxford: Oxford University Press, 2009), loc. 372 of 2485, Kindle.
78. Thomas Davis, *Shifting Sands: The Rise and Fall of Biblical Archaeology* (Oxford: Oxford University Press, 2004), Kindle.
79. Robert Wood, *The Ruins of Palmyra, Otherwise Tedmore in the Desart* (London: Robert Wood, 1753).
80. British Film Institute National Archive, "Ruins of Palmyra and Baalbek," YouTube video, 9:46, posted October 27, 2016, https://www.youtube.com/watch?v=Q2J6IFljEuA.
81. The full statement of significance involved three criteria:

> Criterion (i): The splendour of the ruins of Palmyra, rising out of the Syrian desert north-east of Damascus is testament to the unique

> aesthetic achievement of a wealthy caravan oasis intermittently under the rule of Rome from the Ier to the 3rd century AD. The grand colonnade constitutes a characteristic example of a type of structure which represents a major artistic development. Criterion (ii): Recognition of the splendour of the ruins of Palmyra by travellers in the 17th and 18th centuries contributed greatly to the subsequent revival of classical architectural styles and urban design in the West. Criterion (iv): The grand monumental colonnaded street, open in the centre with covered side passages, and subsidiary cross streets of similar design together with the major public buildings, form an outstanding illustration of architecture and urban layout at the peak of Rome's expansion in and engagement with the East. The great temple of Ba'al is considered one of the most important religious buildings of the 1st century AD in the East and of unique design. The carved sculptural treatment of the monumental archway through which the city is approached from the great temple is an outstanding example of Palmyrene art. The large-scale funerary monuments outside the city walls in the area known as the Valley of the Tombs display distinctive decoration and construction methods.

UNESCO, *Adoption of Retrospective Statements of Outstanding Universal Value*, WHC-10/34.COM/8E (Paris: UNESCO, 2010), 33, https://whc.unesco.org/archive/2010/whc10-34com-8Ee.pdf.

82. See, e.g., Brian Fagan, *Return to Babylon: Travelers, Archaeologists, and Monuments in Mesopotamia* (Boulder: University of Colorado Press, 2007).
83. Neil Asher Silberman, "Promised Lands and Chosen Peoples: The Politics and Poetics of Archaeological Narrative," in *Nationalism, Politics and the Practice of Archaeology*, ed. Philip Kohl and Clare Fawcett (Cambridge: Cambridge University Press, 1996), 255.
84. See, e.g., Trigger, *History of Archaeological Thought*, 256.
85. John Hobson, *The Eastern Origins of Western Civilization* (Cambridge: Cambridge University Press, 2004), 23.
86. Peter Frankopan, *The Silk Roads: A New History of the World* (London: Bloomsbury, 2015).
87. See Rudolph Pfister, *Textiles de Palmyre* (Paris: Les éditions d'art et d'historie, 1934); Rudolph Pfister, *Nouveaux textiles de Palmyre* (Paris: Les éditions d'art et d'historie, 1937); Rudolph Pfister, *Textiles de Palmyre III* (Paris: Les éditions d'art et d'historie, 1940); Marta Żuchowska, "From China to Palmyra: The Value of Silk," in *Światowit: Annual of the Institute of Archaeology of the University of Warsaw* (Warsaw: Institute of Archaeology of the University of Warsaw, 2013), 133–54; and Marta Żuchowska, "'Grape Picking' Silk from Palmyra: A Han Dynasty Chinese Textile with a Hellenistic Decoration Motif," in *Światowit: Annual of the Institute of Archaeology of the University of Warsaw*, ed. Franciszek Setępniowski, Andrzej Macialowiz, Ludwika Jończyk, and Dariusz Szelag (Warsaw: Institute of Archaeology of the University of Warsaw, 2015), 143–62.

88. The invention of radiocarbon dating in the years after World War II was vital in opening up a global history of prehistoric connections linking Africa to East Asia via the Middle East.
89. Green, "Introduction," 26.
90. Green, "Introduction."
91. Eileen Kane, *Russian Hajj: Empire and the Pilgrimage to Mecca* (Ithaca, NY: Cornell University Press, 2015), Kindle.
92. Kane captures the scale of this long-distance travel in the following terms: "Tens of thousands of Muslims made the hajj through Russian lands every year—tsarist subjects as well as those from Persia, Afghanistan, and China—most by way of the Black Sea. Russia's conquests of Muslim lands and peoples, and its mobility revolution, had, in effect, transformed the empire into a crossroads of the global hajj." Kane, *Russian Hajj*, 3.
93. This was the date of the first English-language version of the guide. The German edition was published two years earlier. Karl Baedeker, *Russia, with Teheran, Port Arthur, and Peking: Handbook for Travellers* (Leipzig: Karl Baedeker, 1914).
94. See Frederick Bohrer, *Photography and Archaeology* (London: Reaktion Books, 2011); and Rosalind Morris, *Photographies East: The Camera and Its Histories in East and Southeast Asia* (Durham, NC: Duke University Press, 2009).
95. See, e.g., Judy Bonavia, *Collins Illustrated Guide to the Silk Road* (London: Collins, 1988); and Robert Collins, *East to Cathay: The Silk Road* (New York: McGraw-Hill Book Co., 1986).
96. See A. Gayamov, "Soviet Music," *Soviet Travel* 3 (1933): 8–11; and Boris Olenin, "Sukhum-Kaleh, City of Joy," *Soviet Travel* 3 (1933): 15–17.
97. Elena Sudakova, ed., *See USSR: Intourist Posters and the Marketing of the Soviet Union* (London: GRAD Publishing, 2013), 77.
98. Amilcare Iannucci and John Tulk, "From Alterity to Holism: Cinematic depictions of Marco Polo and His Travels," in *Marco Polo and the Encounter of East and West*, ed. Suzanne Conklin Akbari and Amilcare Iannucci (Toronto: University of Toronto, 2008), 210.
99. Suzanne Akbari, "Introduction: East, West, and In-Between," in *Marco Polo and the Encounter of East and West*, ed. Suzanne Akbari and Amilcare Iannucci (Toronto: University of Toronto Press, 2008), 3–20.
100. Lawrence Thaw and Margaret Thaw, "Along the Old Silk Routes," *National Geographic Magazine* 78 (1940): 453–86.
101. Kuzmina, *Prehistory of the Silk Road*, 5–6.
102. See, e.g., Fagan, *Return to Babylon*; Guha-Thakurta, *Monuments, Objects, Histories*; and Maurizio Peleggi, *Thailand: The Worldly Kingdom* (London: Reaktion, 2007).
103. Michael Clarke, *Xinjiang and China's Rise in Central Asia—A History* (London: Routledge, 2011), 7.
104. Chin, "Invention of the Silk Road, 1877," 217.

105. See Seng Tan and Amitav Acharya, eds., *Bandung Revisited: The Legacy of the 1955 Asian-African Conference for International Order* (Singapore: NUS Press, 2008).
106. Laura Wong, "Relocating East and West: UNESCO's Major Project on the Mutual Appreciation of Eastern and Western Cultural Values," *Journal of World History* 19, no. 3 (2008): 349–74.
107. Wong, "Relocating East and West," 349.
108. Although such themes would today be labeled "tangible" and "intangible" heritage, these terms rarely featured in the project. The concept of heritage was yet to form part of intergovernmental cultural policy discourse, a situation that began to change in the early 1970s through the formulation of the World Heritage Convention.
109. The event lasted from October 28 to November 5 and involved the presentation of more than fifty papers. See the report: Japanese National Commission for UNESCO, ed., *Research in Japan in History of Eastern and Western Cultural Contacts: Its Development and Present Situation* (Tokyo: Japanese National Commission for UNESCO, 1957).
110. Japanese National Commission for UNESCO, *Research in Japan.*
111. Matsuda cites Albert Hermann, rather than Ferdinand von Richthofen, as the source of the term *Seidenstrasse.*
112. Hisao Matsuda, "General Survey: The Development of Researches in the History of the Intercourse between East and West in Japan," in *Research in Japan in History of Eastern and Western Cultural Contacts: Its Development and Present Situation*, ed. Japanese National Commission for UNESCO (Tokyo: Japanese National Commission for UNESCO, 1957), 7.
113. Masao Mori, "The Steppe Route," in *Research in Japan in History of Eastern and Western Cultural Contacts: Its Development and Present Situation*, ed. Japanese National Commission for UNESCO (Tokyo: Japanese National Commission for UNESCO, 1957), 19–34.
114. Namio Egami, "The Silk Road and Japan," in *The Sea Route: The Grand Exhibition of Silk Road Civilizations*, ed. Silk Road Exposition (Nara, Japan: Nara National Museum, 1988), 10–20.
115. Quotes from Silk Road Exposition, ed., *The Route of Buddhist Art: The Grand Exhibition of Silk Road Civilizations* (Nara, Japan: Nara National Museum, 1988), 9, 13; Silk Road Exposition, ed., *The Oasis and Steppe Routes: The Grand Exhibition of Silk Road Civilizations* (Nara, Japan: Nara National Museum, 1988), 9; and Silk Road Exposition, ed., *The Sea Route: The Grand Exhibition of Silk Road Civilizations* (Nara, Japan: Nara National Museum, 1988), 7.
116. For a brief background to this relationship, see "The Project on the Archaeological Research Project on the Sites of Palmyra," Japan Consortium for International Cooperation in Cultural Heritage, http://www.jcic-heritage.jp/en/project/middle_east_palmyra_201109/.
117. Luce Boulnois, *The Silk Road* (London: George Allen and Unwin, 1966).

118. Jan Myrdal, *The Silk Road: A Journey from the High Pamirs and Ili through Sinkiang and Kansu* (London: Victor Gollancz, 1980), 7.
119. The exhibition did not identity a maritime silk road, instead describing various forms of maritime trade and empire across the Indian Ocean and East Asia regions. See John Volmer, E. J. Keall, and E. Nagai-Berthrong, *Silk Roads—China Ships: An Exhibition of East-West Trade* (Toronto: Royal Ontario Museum, 1983).
120. Yu Quiyu, *A Bittersweet Journey through Culture* (New York: CN Times Books, 2015), locs. 1281–86 of 5330, Kindle.
121. For further details, see "The International Dunhuang Project: The Silk Road Online," British Library, http://idp.bl.uk.
122. The full title of the initiative was the Integral Study of the Silk Roads: Roads of Dialogue. For further details of the project's aims and scope, see John Lawton, ed., *The Integral Study of the Silk Roads: Roads of Dialogue* (Paris: UNESCO, 2008).
123. Federico Mayor, "Preface of the Director-General of UNESCO," in *The Integral Study of the Silk Roads: Roads of Dialogue*, ed. John Lawton (Paris: UNESCO, 2008), 3.
124. For a list of the project's scholarly events, see Vadime Elisseef, *The Silk Roads: Highways of Culture and Commerce* (New York: Berghahn Books, 2000), appendix 1, 329–32.
125. See Elisseef, *Silk Roads*.
126. Joseph Ting, *The Maritime Silk Route: 2000 Years of the South China Sea* (Hong Kong: Urban Council, 1996).
127. Smithsonian Folklife Festival, "The Silk Road Ensemble," https://festival.si.edu/2002/the-silk-road/the-silk-road-ensemble/smithsonian.
128. Richard Kennedy, "The Silk Road: Connecting Cultures, Creating Trust," Smithsonian Folklife Festival, https://festival.si.edu/2002/the-silk-road/the-silk-road-connecting-cultures-creating-trust/smithsonian.
129. Finbarr Flood and Gülru Necipoğlu, *A Companion to Islamic Art and Architecture* (New York: John Wiley and Sons, 2017), loc. 2537 of 3724, Kindle.
130. See Ulf Hannerz, "Geocultural Scenarios," in *Frontiers of Sociology*, ed. Peter Hedstrom and Bjorn Wittrock (Leiden: Brill, 2009), 267–88.
131. Żuchowska, "'Grape Picking' Silk from Palmyra," 143–62.
132. UNESCO TV and NHK Nippon Hoso Kyokai, "UNESCO/NHK Videos on Heritage," UNESCO, https://whc.unesco.org/en/list/23/video.
133. GovTrack, "H.R. 2867—105th Congress: Silk Road Strategy Act of 1997," https://www.govtrack.us/congress/bills/105/hr2867.
134. S. Frederick Starr, introduction to *The New Silk Roads: Transport and Trade in Greater Central Asia*, ed. S. Frederick Starr (Washington, DC: Central Asia-Caucasus Institute, 2007), 5.

CHAPTER THREE

1. Vasillis Trigkas, "The New Silk Road: Cultural and Economic Diplomacy China-Greece-Europe," Tsinghua University, http://www.imir.tsinghua.edu.cn/publish/iisen/7259/2014/20141010134153540200352/20141010134153540200352_.html.
2. Trigkas, "New Silk Road."
3. Trigkas, "New Silk Road."
4. Trigkas, "New Silk Road."
5. Xi Jinping, "Promote Friendship between Our People and Work Together to Build a Bright Future" (speech, Nazabayev University, Astana, September 7, 2013), http://www.fmprc.gov.cn/mfa_eng/wjdt_665385/zyjh_665391/t1078088.shtml.
6. Xi Jinping, "Speech by Chinese President Xi Jinping to Indonesian Parliament" (speech, Jakarta, October 2, 2013), http://www.asean-china-center.org/english/2013–10/03/c_133062675.htm.
7. Xi Jinping, "Speech by Chinese President Xi Jinping to Indonesian Parliament."
8. Xi Jinping, "Promote Friendship between Our People."
9. See, e.g., Wu Xiaogang and Xi Song, "Ethnic Stratification amid China's Economic Transition: Evidence from the Xinjiang Uyghur Autonomous Region," *Social Science Research* 44 (2017): 158–72; and Roy Rogers and Jatswan Sidhu, "International Norms and Human Rights Conditions in the Xinjiang Uyghur Autonomous Region (XUAR)," *Malaysian Journal of International Relations* 4 (2017): 109–37.
10. Qu Zhe, "Speech on the Belt and Road Initiative by Chinese Ambassador Qu Zhe" (speech, Tallinn, Estonia, April 13, 2013), http://www.fmprc.gov.cn/mfa_eng/wjb_663304/zwjg_665342/zwbd_665378/t1254259.shtml.
11. Roger Irvine, "Nationalists or Internationalists? China's International Relations Experts Debate the Future," *Journal of Contemporary China* 26, no. 106 (2017): 587; and Zhu Zhiqun, *China's New Diplomacy: Rationale, Strategies and Significance* (Farnham, UK: Ashgate, 2013), Kindle.
12. Zhu Zhiqun, *China's New Diplomacy*, loc. 3007.
13. Emily Yeh, "Introduction: The Geoeconomics and Geopolitics of Chinese Development and Investment in Asia," *Eurasian Geography and Economics* 57, no. 3 (2016): 275–85.
14. Weidong Sun, "Peaceful Development and Win-Win Cooperation" (speech, National Defence University, Islamabad, Pakistan, September 12, 2013), http://pk.china-embassy.org/eng/zbgx/t1076545.htm.
15. Peter Ferdinand, "Westward Ho—The China Dream and 'One Belt, One Road': Chinese Foreign Policy under Xi Jinping," *International Affairs* 92, no. 4 (2016): 950.
16. Qu Zhe, "Speech on the Belt and Road Initiative by Chinese Ambassador Qu Zhe."

17. Yang Jiechi, "Jointly Undertake the Great Initiatives with Confidence and Mutual Trust" (speech, Boao, Hainan, April 10, 2014), http://www.fmprc.gov.cn/mfa_eng/wjdt_665385/zyjh_665391/t1145860.shtml.
18. Chris Devonshire-Ellis, *China's New Economic Silk Road: The Great Eurasian Game and the String of Pearls* (Hong Kong: Asia Briefing, Dezan Shira and Associates, 2015).
19. Kent Calder, *The New Continentalism: Energy and Twenty-First-Century Eurasian Geopolitics* (New Haven, CT: Yale University Press, 2012), 161.
20. Yang Jiechi, "Jointly Undertake the Great Initiatives with Confidence and Mutual Trust."
21. Xi Jinping, "Speech by Chinese President Xi Jinping to Indonesian Parliament." See also Xi Jinping's keynote speech, "A New Blueprint for Global Economic Growth," at the opening ceremony of the G20 Summit, 2016: "We need to rely on partnerships, uphold the vision of win-win results, step up practical cooperation across the board, continue to enrich and expand cooperation and ensure that such cooperation delivers outcomes to meet people's expectations. We need to enable people of different countries, cultures and historical backgrounds to deepen exchanges, enhance mutual understanding and jointly build a community of shared future for mankind" (Hangzhou, September 4, 2016), http://www.globalresearch.ca/chinas-president-xi-jinpings-opening-address-of-g20-summit-a-new-blueprint-for-global-economic-growth/5543895.
22. Zhang Dejiang, "Keynote Speech of China's Top Legislator at Belt and Road Summit in Hong Kong" (speech, Hong Kong, May 18, 2016), http://news.xinhuanet.com/english/china/2016–05/18/c_135368795.htm.
23. Yang Jiechi, "Jointly Undertake the Great Initiatives with Confidence and Mutual Trust."
24. Jiang Zemin, "Enhance Mutual Understanding and Build Stronger Ties of Friendship and Cooperation" (speech, Harvard University, Cambridge, MA, November 1, 1997), http://www.china-embassy.org/eng/zmgx/zysj/jzxfm/t36252.htm.
25. Wen Jiabao, "See China in the Light of Her Development" (speech, University of Cambridge, February 2, 2009), http://news.xinhuanet.com/english/2009-02/03/content_10753336_2.htm. See also the following: Hu Jintao, "Speech by Chinese President Hu Jintao Given to Federal Parliament" (speech, Canberra, October 24, 2003), http://www.smh.com.au/articles/2003/10/24/1066631618612.html; Wen Jiabao, "Chinese Premier Wen Jiabao's Speech at the Arab League" (speech, Cairo, November 7, 2009), http://en.people.cn/90001/90776/90883/6806808.html; and Wen Jiabao, "Strengthen Good-Neighbourly Relations and Deepen Mutually Beneficial Cooperation" (speech, Balai Kartini, Indonesia, April 30, 2011), http://news.xinhuanet.com/english2010/china/2011-05/01/c_13853424.htm.
26. Yao Jianing, ed., "PLA Navy Training Ship *Zhenghe* Embarks on Far-Sea Voyage," *China Military*, October 25, 2016, http://english.chinamil.com.cn/view/2016-10/25/content_7324566.htm.

27. See Gregory Poling, *Grappling with the South China Sea Policy Challenge* (Washington, DC: Center for Strategic and International Studies, 2015); International Crisis Group, *Stirring Up the South China Sea (II): Regional Responses* (Brussels: International Crisis Group, 2012); Zhao Hong, "The South China Sea Dispute and China-ASEAN Relations," *Asian Affairs* 44, no. 1 (2013): 27–43; and Aminuk Karim Mohd, "The South China Sea Disputes: Is High Politics Overtaking?," *Pacific Focus* 28, no. 1 (2013): 99–119.
28. See, e.g., Nguyen-Dang Thang and Hong Thao Nguyen, "China's Nine Dotted Lines in the South China Sea: The 2011 Exchange of Diplomatic Notes between the Philippines and China," *Ocean Development and International Law* 43, no. 1 (2012): 35–56; and Hong Thao Nguyen, "Vietnam's Position on the Sovereignty over the Paracels and the Spratlys: Its Maritime Claims," *Journal of East Asia and International Law* 5, no. 1 (2012): 165–211.
29. Ross Babbage, *Countering China's Adventurism in the South China Sea: Strategy Options for the Trump Administration* (Washington, DC: Center for Strategic and Budgetary Assessments, 2016), http://csbaonline.org/research/publications/countering-chinas-adventurism-in-the-south-china-sea-strategy-options-for-t.
30. Xi Jinping, "Forging a Strong Partnership to Enhance Prosperity in Asia" (speech, Singapore, November 7, 2015), http://learning.sohu.com/20151113/n426405360.shtml.
31. See Lim Tai Wei, introduction to *China's One Belt One Road Initiative*, ed. Lim Tai Wei, Henry Chan Hing Lee, Katherine Tseng Hui-Yi, and Lim Wen Xi (London: Imperial College Press, 2016), 3–18.
32. Yang Jiechi, "Jointly Undertake the Great Initiatives with Confidence and Mutual Trust."
33. According to Wang Yi:

> The "Belt and Road" initiative will be a public good China provides to the world, and we welcome all countries, international organizations, multinationals, financial institutions and [nongovernmental organizations] to join in and be part of the specific cooperation programs. On the security front, we need to commit ourselves to building a world of security for all. This year marks the 70th anniversary of the end of World War II. For our world, peace has not come easily and must be preserved however hard we need to work. We need to better leverage the safeguard mechanisms established by the UN Charter to prevent war and uphold peace. . . . China is an active contributor to international peace and security operations. It has sent nearly 30,000 men and women to UN peacekeeping operations, more than other permanent members of the UN Security Council. By now, 19 Chinese fleets have conducted escort missions for the safe passage of over 5,800 ships in the Gulf of Aden and off the coast of Somalia.

Wang Yi, "Toward a New Type of International Win-Win Cooperation" (speech, Beijing, March 25, 2015), http://www.fmprc.gov.cn/mfa_eng/

zxxx_662805/t1248487.shtml?utm_campaign=buffer&utm_content=buffer1e863&utm_medium=social&utm_source=twitter.com.

34. During his launch speech of the Maritime Silk Road in Indonesia, Xi Jinping elaborated on China's assistance in the aftermath of the Aceh tsunami in 2004. See Xi Jinping, "Speech by Chinese President Xi Jinping to Indonesian Parliament."
35. Irvine, "Nationalists or Internationalists?," 2–3.
36. Irvine, "Nationalists or Internationalists?," 1.
37. Irvine, "Nationalists or Internationalists?," 11.
38. Xi Jinping's quotes are from "President Xi Jinping Delivers Important Speech at Pakistan's Parliament Entitled 'Building a China-Pakistan Community of Shared Destiny to Pursue Closer Win-Win Cooperation,'" Ministry of Foreign Affairs of the People's Republic of China, April 21, 2015, http://www.fmprc.gov.cn/mfa_eng/topics_665678/xjpdbjstjxgsfwbfydnxycxyfldrhyhwlhy60znjnhd/t1257288.shtml; and "Speech by Chinese President Xi Jinping to Indonesian Parliament." Wang Yi's quote is from "Toward a New Type of International Win-win Cooperation."
39. See, e.g., discussions over Tehran appreciating China's ongoing engagement during a period of sanctions—"China, Iran Vow Tighter Ties as Xi Visits," *Channel NewsAsia*, January 23, 2016, http://www.channelnewsasia.com/news/asiapacific/china-iran-vow-tighter/2451404.html. For an appreciation of Serbia and China's shared fight "against aggression and oppression," see Wang Huijuan and Nemanja Cabric, "Interview: Belt and Road Initiative to Foster China-Serbia Win-Win Cooperation: Serbian Expert," *Xinhuanet*, August 14, 2016, http://news.xinhuanet.com/english/2016-08/14/c_135596795.htm.
40. Ambassador Prasad Kariyawasam, "Sri Lanka—A Hub in the Indian Ocean" (speech, Honolulu, HI, February 11, 2016), http://slembassyusa.org/embassy_press_releases/remarks-by-ambassador-prasad-kariyawasam-at-the-east-west-center-hawaii-sri-lanka-a-hub-in-the-indian-ocean/.
41. See Mohamed Azmin Ali, "The Maritime Silk Road and a New Paradigm for Globalisation" (speech, Malaysia, November 7, 2015), http://azminali.com/speech-the-5th-malaysia-china-entrepreneur-conference/.
42. For example, according to Irakli Garibashvili, prime minister of Georgia, in September 2015: "This year, we have restored the fast rail link between China and Georgia, making it the fastest land route leading to Europe. This is the first time since 1513, when the Portuguese Karavelas reached the Chinese coast, that the route is operational. Today, the process of the Silk Road's revival is also creating a closer relationship between our countries. Hopefully, not only trade, but also people, ideas, knowledge and culture will enrich our relations." "PM Discusses Georgia-China Partnership at Beijing University," Government of Georgia, September 11, 2015, http://gov.ge/index.php?lang_id=ENG&sec_id=412&info_id=51429.

In "Speech by Special Assistant to the Prime Minister Syed Tariq Fatemi at the Karachi Council of Foreign Relations" (speech, Karachi, February 2, 2015, http://www.mofa.gov.pk/chile/pr-details.php?prID=2546), Syed Tariq Fatemi, special assistant to the prime minister of Pakistan, said:

> History shows that inter-regional and intra-regional trade has been one of the driving forces in ushering in prosperity and progress of mankind. Besides enhancing prospects for material well-being, the trade routes have played a crucial role in facilitating exchange of ideas and deepening of civilizational links. The ancient Silk Road, one of the most famous of such routes, was a great channel through which scores of countries exchanged goods, cultural traditions, and creative ideas. It played a unique role in promoting foreign trade and fostering political and cultural ties in Asia and beyond. The advent of modern technologies and development of alternate routes gradually pushed this historical route into relative disuse. But the evolving geo-political realities and growing primacy of economic priorities has catapulted this relic of the past into a modern-day imperative.
>
> One of the earliest steps taken by the Prime Minister, after taking office in June 2013, was to conclude the MoU on establishing the China-Pakistan Economic Corridor (CPEC) that would connect Pakistan with China's Western regions, through a network of roads, railways and fiber-optic linkages. Pakistan and China see the Corridor as a potential catalyst for regional economic integration—besides fostering regional harmony and closer relations between China and its neighbours.

43. Yang Jiechi, "Jointly Undertake the Great Initiatives with Confidence and Mutual Trust."

CHAPTER FOUR

1. "Vision and Actions on Jointly Building Silk Road Economic Belt and 21st-Century Maritime Silk Road," National Development and Reform Commission, Ministry of Foreign Affairs, and Ministry of Commerce of the People's Republic of China, March 28, 2015, http://en.ndrc.gov.cn/newsrelease/201503/t20150330_669367.html.
2. "Silk Road Dunhuang International Music Forum Starts," Asia Pacific Exchange and Cooperation Foundation, July 26, 2015, http://www.apecf.org/en/foundationnews/20150726.html.
3. Divia Harilela, "Hong Kong Fashion Designer Vivienne Tam Talks about Her First Public Show of Art," *South China Morning Post*, April 15, 2016, http://www.scmp.com/lifestyle/fashion-luxury/article/1936128/hong-kong-fashion-designer-vivienne-tam-talks-about-her.

4. Wang Kaihao and Hu Meidong, "Arts Festival Reflects Ancient Silk Road Cultures," State Council, People's Republic of China, November 9, 2015, http://english.gov.cn/news/photos/2015/11/09/content_281475231256090.htm.
5. See, e.g., the following short report: UNESCO, *Integral Study of the Silk Roads: Roads of Dialogue, 1998–1997* (Paris: UNESCO, 2008), http://unesdoc.unesco.org/images/0015/001591/159189E.pdf.
6. This was recognized during a meeting of experts in Ittingen, Switzerland, that focused on the challenges of listing and the transnational management of serial properties. See UNESCO, *Report on Serial Nominations and Properties* (Paris: UNESCO, 2010), http://whc.unesco.org/archive/2010/whc10-34Com-9Be.pdf.
7. For further details, see Tim Winter, "Heritage Diplomacy: Entangled Materialities of International Relations," *Future Anterior* 21, no. 1 (2016): 16–34.
8. See UNESCO and Republic of Korea Funds-in-Trust-Project, *Support for the Preparation for the World Heritage Serial Nomination of the Silk Roads in South Asia* (Paris: UNESCO, 2016), http://unesdoc.unesco.org/images/0024/002460/246096e.pdf.
9. See UNESCO and Republic of Korea Funds-in-Trust-Project, *Support for the Preparation for the World Heritage Serial Nomination of the Silk Roads in South Asia*, 2.
10. Tim Williams, *The Silk Roads: An ICOMOS Thematic Study* (Paris: International Council of Monuments and Sites, 2014).
11. As of late 2015, the membership of the committee was made up of Afghanistan, China, India, Iran, Japan, Kazakhstan, Kyrgyzstan, Nepal, Pakistan, Republic of Korea, Tajikistan, Turkey, Turkmenistan, and Uzbekistan.
12. This included the referred Penjikent-Samarkand-Poykent Corridor and two new nominations: the Fergana-Syrdarya Corridor, a collaboration between Tajikistan and Uzbekistan, and the South Asian Silk Road Corridor, administered by Bhutan, China, India, and Nepal. Key meetings here include International Forum on the "Great Silk Roads," Almaty, October 14–16, 2013, and Fourth Meeting of the Coordinating Committee on the Serial World Heritage Nomination of the Silk Roads, Almaty, November 24–25.
13. The government of Japan's donations were $985,073 for 2011–2014 and $697,796 for 2015–2018. See UNESCO, "Silk Roads World Heritage Serial and Transnational Nomination in Central Asia: A UNESCO/Japanese Funds-in-Trust Project," http://whc.unesco.org/en/activities/825/; and UNESCO, "Support for the Silk Roads World Heritage Sites in Central Asia (Phase II)," http://whc.unesco.org/en/activities/870. For South Korea, see UNESCO and Republic of Korea Funds-in-Trust-Project, *Support for the Preparation for the World Heritage Serial Nomination of the Silk Roads in South Asia*.

14. See UNESCO Kazakhstan, "UNESCO and the Centre for the Rapprochement of Cultures in Kazakhstan Have Defined an Action Plan for the Intercultural Dialogue," August 5, 2016, http://en.unesco.kz/the-first-scientific-and-methodological-meeting-of-the-council-of-the-state-museum-of-the. In August 2016, for example, the Centre co-organized a conference, the New Silk Way as a Dialogue of Cultures, with participants from China and Kazakhstan.
15. See, e.g., Ananth Krishnan, "China Offers to Develop Chittagong Port," *The Hindu*, March 15, 2010, http://www.thehindu.com/news/international/china-offers-to-develop-chittagong-port/article245961.ece; and Shannon Tiezzi, "Chinese Company Wins Contract for Deep Sea Port in Myanmar," *The Diplomat*, January 1, 2016, http://thediplomat.com/2016/01/chinese-company-wins-contract-for-deep-sea-port-in-myanmar/.
16. Talha Bin Habib, "China to Help Build 2nd Runway, 3rd Terminal at HSIA," *Financial Express*, July 4, 2013, http://print.thefinancialexpress-bd.com/old/index.php?ref=MjBfMDdfMDRfMTNfMV85MF8xNzUyNzU=; and "China to Build 2 More Bridges in Bangladesh," *Daily Star*, August 26, 2015, http://www.thedailystar.net/country/china-build-2-more-bridges-bangladesh-132925.
17. Mahwish Chowdhary, "China's Billion-Dollar Gateway to the Subcontinent: Pakistan May Be Opening a Door It Cannot Close," *Forbes*, August 25, 2015, http://www.forbes.com/sites/realspin/2015/08/25/china-looks-to-pakistan-to-expand-its-influence-in-asia/#11189f987179.
18. Zafar Bhutta, "Chinese Offer to Finance Whole $2bn LNG Project," *Express Tribune*, May 14, 2016, http://tribune.com.pk/story/1103009/lng-terminal-and-pipeline-chinese-offer-to-finance-whole-2-billion-project/; and Chowdhary, "China's Billion-Dollar Gateway."
19. See Vladislav Inozemtsev, "Why Kazakhstan Holds the Keys to the Global Economy," *The Independent*, November 10, 2015, http://www.independent.co.uk/voices/why-kazakhstan-holds-the-keys-to-the-global-economy-a6727391.html; and Matthew Kiernan, "Silk Road Key to China's Next Move," *Huffington Post*, August 31, 2015, http://www.huffingtonpost.com/matthew-j-kiernan/silk-road-key-to-chinas-n_b_8068318.html.
20. AKIPress, "Kazakh-Chinese Business Council Sign Deals Worth $10bn," *Eurasian Business Briefing*, December 15, 2015, http://www.eurasianbusinessbriefing.com/10bn-worth-of-deals-signed-at-kazakh-chinese-business-council/.
21. "China Lends $300m to Build Strategic Road in Kyrgyzstan," Alexander's Gas & Oil Connections: An Institute for Global Energy Research, June 16, 2015, http://www.gasandoil.com/oilaround/2015/06/china-lends-300-mm-to-build-strategic-road-in-kyrgyzstan.
22. Sudha Ramachandran, "Iran, China and the Silk Road Train," *The Diplomat*, March 30, 2016, http://thediplomat.com/2016/03/iran-china-and-the-silk-road-train/.

23. Saeed Shah, "China to Build Pipeline from Iran to Pakistan," *Wall Street Journal*, April 9, 2015, http://www.wsj.com/articles/china-to-build-pipeline-from-iran-to-pakistan-1428515277.
24. Afshin Majlesi, "Tehran Hosts Sino-Iranian Friendship Conference as Ancient Silk Road Backs to Life," *Tehran Times*, October 24, 2016, http://www.tehrantimes.com/news/407630/Tehran-hosts-Sino-Iranian-friendship-conference-as-ancient-Silk.
25. Williams, *Silk Roads*, 32, 71–73.
26. Ministry of Foreign Affairs, Islamic Republic of Afghanistan, *The Sixth Regional Economic Cooperation Conference on Afghanistan Report* (n.p.: Ministry of Foreign Affairs, Islamic Republic of Afghanistan, 2015), 10.
27. UN World Tourism Organization, *Silk Road Action Plan 2016/7* (Madrid: UN World Tourism Organization, 2017), http://silkroad.unwto.org/node/30284.
28. Efforts by the World Tourism Organization to develop the overland Silk Road tourism sector stretch back to the early 1990s, when it gathered support from a number of countries across Central and East Asia through a series of declarations and symposia. For an overview, see "WTO and the Silk Road," Manzara Tourism, accessed December 22, 2016, http://www.manzaratourism.com/gsr_wto.
29. UN World Tourism Organization, *Silk Road Action Plan 2016/7.*
30. See Tang West Market Group, "TWMG's Contribution to the Revival of the Silk Road of UNESCO," September 2015, accessed February 13, 2016, http://www.tangwestmarket.com/2015/09/twmgs-contribution-to-the-revival-of-the-silk-road-of-unesco-3/.
31. Atul Aneja, "China Developing Soft-Power Infra along Silk Road," *The Hindu*, October 18, 2016, http://www.thehindu.com/news/international/China-developing-soft-power-infra-along-Silk-Road/article14308346.ece.
32. See Emily Yeh and Elizabeth Wharton, "Going West and Going Out: Discourses, Migrants, and Models in Chinese Development," *Eurasian Geography and Economics* 57, no. 3 (2016): 286–315.
33. See, e.g., Chowdhary, "China's Billion-Dollar Gateway."
34. See, e.g., Roy Rogers and Jatswan Sidhu, "International Norms and Human Rights Conditions in the Xinjiang Uyghur Autonomous Region (XUAR)," *Malaysian Journal of International Relations* 4 (2017): 109–37; S. Frederick Starr, ed., *Xinjiang: China's Muslim borderland* (Armonk, NY: M. E. Sharpe, 2004); and Wu Xiaogang and Xi Song, "Ethnic Stratification amid China's Economic Transition: Evidence from the Xinjiang Uyghur Autonomous Region," *Social Science Research* 44 (2014): 158–72.
35. See Dru Gladney, "The Party-State's Nationalist Strategy to Control the Uyghur: Silenced Voices," in *Routledge Handbook of the Chinese Communist Party*, ed. Willy Wo-Lap Lam (Oxford: Routledge, 2018), p. 325, Kindle.

36. For further details, see James Leibold, "Interethnic Conflict in the PRC: Xinjiang and Tibet as Exceptions?," in *Protest and Xinjiang: Unrest in China's West*, ed. Ben Hillman and Gray Tuttle (New York: Columbia University Press, 2016), 223–50.
37. Michael Clarke, "Xinjiang from the 'Outside-In' and the 'Inside-Out': Exploring the Imagined Geopolitics of a Contested Region," in *Inside Xinjiang: Space, Place and Power in China's Muslim Far Northwest*, ed. Anna Hayes and Michael Clarke (Oxford: Routledge, 2016), 227.
38. James Leibold, *Reconfiguring Chinese Nationalism: How the Qing Frontier and Its Indigenes Became Chinese* (New York: Palgrave Macmillan, 2007), Kindle. See also James Millward, *Eurasian Crossroads: A History of Xinjiang* (London: Hurst and Co., 2007).
39. Limited space here precludes a longer discussion of events in Xinjiang in the twentieth century. For a more detailed account of the region, see Justin Jacobs, *Xinjiang and the Modern Chinese State (Studies on Ethnic Groups in China)* (Seattle: University of Washington Press, 2016), Kindle; and Millward, *Eurasian Crossroads*, chaps. 5–7.
40. For a more detailed account of the period, see Clarke, *Xinjiang and China's Rise in Central Asia*, 1–15.
41. For a more detailed discussion, see Jacobs, *Xinjiang and the Modern Chinese State*, 169–94.
42. Clarke, "Xinjiang from the 'Outside-In' and the 'Inside-Out,'" 239.
43. See, e.g., Michael Clarke, "China and the Uyghur: The 'Palestinization' of Xinjiang," *Middle East Policy* 22, no. 3 (2015): 127–46; Shale Horowitz and Peng Yu, "Holding China's West: Explaining CCP Strategies of Rule in Tibet and Xinjiang," *Journal of Chinese Political Science* 20, no. 4 (2015): 451–75; and Angel Ryono and Matthew Galway, "Xinjiang under China: Reflections on the Multiple Dimensions of the 2009 Urumqi Uprising," *Asian Ethnicity* 16, no. 2 (2015): 235–55.
44. James Millward, "Uyghur Art Music and the Ambiguities of Chinese Silk Roadism in Xinjiang," *Silkroad Foundation Newsletter*, June 2005, http://www.silkroadfoundation.org/newsletter/v013num1/3_uyghur.php.
45. Millward, "Uyghur Art Music and the Ambiguities of Chinese Silk Roadism in Xinjiang."
46. Clarke, *Xinjiang and China's Rise in Central Asia*, 5.
47. Rian Thum, "How Should the World Respond to Intensifying Repression in Xinjiang?," *ChinaFile*, June 4, 2018, http://www.chinafile.com/conversation/how-should-world-respond-intensifying-repression-xinjiang.
48. James Millward, "What It's Like to Live in a Surveillance State," *New York Times*, February 3, 2018, https://www.nytimes.com/2018/02/03/opinion/sunday/china-surveillance-state-uighurs.html.
49. See, e.g., Leakthina Chau-Pech Ollier and Tim Winter, eds., *Expressions of Cambodia: The Politics of Tradition, Identity and Change* (London: Routledge,

2006); and Shelly Errington, *The Death of Authentic Primitive Art and Other Tales of Progress* (Berkeley: University of California Press, 1998).

50. "Vision and Actions on Jointly Building Silk Road Economic Belt and 21st-Century Maritime Silk Road."
51. Jeff Adams, "The Role of Underwater Archaeology in Framing and Facilitating the Chinese National Strategic Agenda," in *Cultural Heritage Politics in China*, ed. Tami Blumenfield and Helaine Silverman (New York: Springer, 2013), 274.
52. Adams, "Role of Underwater Archaeology," 274.
53. Adams, "Role of Underwater Archaeology," 267.
54. For further details, see China Underwater Cultural Heritage, *China: National Conservation Center for Underwater Cultural Heritage*, http://www.unesco.org/new/fileadmin/MULTIMEDIA/HQ/CLT/pdf/CUCH_brochure_en (2)_02.pdf.
55. These included the *Sandaogang* off the Liaoning coast (1992–1997), *Wanjiao-01* and *Daliandao* off the Fujian coast (2005–2007), and the *Huaguangjiao* wreck found in the Paracels (2007–2008). See Andrew Erickson and Kevin Bond, "Archaeology and the South China Sea," *The Diplomat*, July 20, 2015, http://thediplomat.com/2015/07/archaeology-and-the-south-china-sea/.
56. Adams, "Role of Underwater Archaeology," 271.
57. For a comprehensive overview of the publications on Zheng He across several languages, see Ying Liu, Zhongping Chen, and Gregory Blue, *Zheng He's Maritime Voyages (1405–1433) and China's Relations with the Indian Ocean World: A Multilingual Bibliography* (Leiden: Brill, 2014).
58. For an example of this discourse, see Li Rongxia, "'Significance of Zheng He's Voyages,'" *Beijing Review*, 2005, http://www.bjreview.cn/EN/En-2005/05-28-e/china-2.htm.
59. For further details of the reconstruction project, see Mara Hvistendahl, "Rebuilding a Treasure Ship," *Archaeology* 61, no. 2 (2008): 40–45.
60. See Bob Wekesa, "Admiral Zheng He and the Diplomatic Value of China's Ancient East African Contacts," Africa-China Reporting Project, July 9, 2013, http://china-africa-reporting.co.za/2013/07/admiral-zheng-he-and-the-diplomatic-value-of-chinas-ancient-east-african-contacts/.
61. See, e.g., "Kenya Islands Show Footprint of Chinese Expeditionary Voyages," *Coastweek.com*, http://www.coastweek.com/3828-Kenyan-islands-with-footprint-of-Chinese-expeditionary-voyages.htm; and Sola Rey, "DNA Test Proves Some East Africans are Descendants of Chinese Sailors Shipwrecked on Kenya's Shores 600 Years Ago," *Sola Rey* (blog), September 17, 2016, http://solarey.net/dna-test-proves-east-africans-descendants-chinese-sailors-shipwrecked-kenyas-shores-600-years-ago/. See also Zoe Murphy, "Zheng He: Symbol of China's 'Peaceful Rise,'" *BBC News*, July 28, 2010, http://www.bbc.com/news/world-asia-pacific-10767321.

62. Megan Gannon, "600 Year Old Coin Found in China," *LiveScience*, March 14, 2013, http://www.livescience.com/27890-chinese-coin-found-in-kenya.html, quoting Chapurukha M. Kusimba of the Field Museum in Chicago.
63. For further details, see Institute of Developing Economies—Japan External Trade Organization, "China in Africa," http://www.ide.go.jp/English/Data/Africa_file/Manualreport/cia_10.html.
64. Chris Devonshire-Ellis, *China's New Economic Silk Road: The Great Eurasian Game and the String of Pearls* (Hong Kong: Asia Briefing, Dezan Shira and Associates, 2015).
65. Geoffrey York, "Africa's Ambitious Mega-Port Project Shrouded in Skepticism," *Globe and Mail*, September 10, 2012, http://www.theglobeandmail.com/news/world/worldview/africas-ambitious-mega-port-project-shrouded-in-skepticism/article551398/.
66. PSCU, "Kenya and China Sign Sh42bn Lamu Port Deal," *Daily Nation*, August 1, 2014, http://www.nation.co.ke/counties/nairobi/Kenya-and-China-Lamu-Port-deal/1954174-2405458-9yviq5/index.html.
67. AidData, "CRBC Builds Railway Link from Mombasa to Nairobi—Concessional Loan (Linked to Project ID #31777)," accessed September 21, 2016, http://china.aiddata.org/projects/35232?iframe=y.
68. Adams, "Role of Underwater Archaeology," 272; and Nalaka Godahewa, "Commemorating Zheng He, the Greatest Navigator to Visit Sri Lanka from China," *Daily FT*, February 4, 2012, http://www.ft.lk/article/69932/Commemorating-Zheng-He—the-greatest-navigator-to-visit-Sri-Lanka-from-China.
69. Ministry of Foreign Affairs, Sri Lanka, "Joint Statement between the People's Republic of China and the Democratic Socialist Republic of Sri Lanka at the Conclusion of the Official Visit of Prime Minister Ranil Wickremesinghe," April 9, 2016, https://www.mfa.gov.lk/jointstatement-slpmvisitchina/.
70. Anusha Ondaatjie, "China Maritime Silk Road Is Sri Lanka's Boon as Xi Visits," *Bloomberg*, September 17, 2014, https://www.bloomberg.com/news/articles/2014-09-15/china-maritime-silk-road-proves-boon-for-sri-lanka-as-xi-arrives.
71. Bharatha Mallawarachi, "Sri Lanka Signs on to China's Maritime 'Silk Road,'" *Sydney Morning Herald*, September 17, 2014, http://www.smh.com.au/world/sri-lanka-signs-on-to-chinas-maritime-silk-road-20140916-10hx4a.html.
72. For further details, see Wade Shepard, "China's Jewel in the Heart of the Indian Ocean," *The Diplomat*, May 9, 2016, http://thediplomat.com/2016/05/chinas-jewel-in-the-heart-of-the-indian-ocean/.
73. The BRICS countries are Brazil, Russia, India, China, and South Africa. For further details, see "Xi Jinping Meets with President Maithripala Sirisena of Sri Lanka," Ministry of Foreign Affairs of the People's Republic

of China, October 17, 2016, https://www.fmprc.gov.cn/mfa_eng/zxxx_662805/t1406780.shtml.

74. Geoff Wade, "The Zheng He Voyages: A Reassessment" (ARI Working Paper No. 31, Asia Research Institute, October 2004), 13.
75. Wade, "Zheng He Voyages," 13.
76. Tan Ta Sen, "Cheng Ho's Guanchang Site in Melaka," in *Zheng He and the Afro-Asian World*, ed. Chia Lin Sien and Sally Kathryn Church (Melaka: Perbadanan Muzium Melaka and International Zheng He Society, 2012), 201.
77. Tan Ta Sen, "Cheng Ho's Guanchang Site in Melaka," 212.
78. Tan Ta Sen, "Cheng Ho's Guanchang Site in Melaka," 212.
79. Amy Chew, "China, Malaysia Tout New 'Port Alliance' to Reduce Customs Bottlenecks and Boost Trade," *South China Morning Post*, April 9, 2016, http://www.scmp.com/news/asia/southeast-asia/article/1934839/china-malaysia-tout-new-port-alliance-reduce-customs; and Robin Bromby, "Down the Maritime Silk Road," *The Australian*, December 6, 2013, http://www.theaustralian.com.au/business/in-depth/down-the-maritime-silk-road/story-fnjy4qn5-1226776242929.
80. Johanes Herlijanto, "Cultivating the Past, Imagining the Future: Enthusiasm for Zheng He in Contemporary Indonesia," in *Zheng He and the Afro-Asian World*, ed. Chia Lin Sien and Sally Kathryn Church (Melaka: Perbadanan Muzium Melaka and International Zheng He Society, 2012).
81. Herlijanto, "Cultivating the Past, Imagining the Future," 141–44.
82. Choirul Mahfud, "Lessons from Zheng He: Love of Peace and Multiculturalism," in *Zheng He and the Afro-Asian World*, ed. Chia Lin Sien and Sally Kathryn Church (Melaka: Perbadanan Muzium Melaka and International Zheng He Society, 2012), 188–91.
83. Xie Feng, "Partnership to Scale New Heights," *China Daily Asia*, April 17, 2015, http://www.chinadailyasia.com/asiaweekly/2015-04/17/content_15253195.html.
84. For further details, see Johanes Herlijanto, "What Does Indonesia's Pribumi Elite Think of Ethnic Chinese Today?," *ISEAS Perspective* 2016, no. 32 (2016): 1–9.
85. Choirul Mahfud, "The Role of Cheng Ho Mosque: The New Silk Road, Indonesia-China Relations in Islamic Cultural Identity," *Journal of Indonesian Islam* 8, no. 1 (2014): 31–34.
86. Ministry of Foreign Affairs of the People's Republic of China, "Joint Statement on Strengthening Comprehensive Strategic Partnership between the People's Republic of China and the Republic of Indonesia," March 27, 2015, http://www.fmprc.gov.cn/mfa_eng/wjdt_665385/2649_665393/t1249201.shtml.
87. Yang Jiechi, "Jointly Build the 21st Century Maritime Silk Road by Deepening Mutual Trust and Enhancing Connectivity" (speech, Beijing, March 28, 2015), http://www.fmprc.gov.cn/mfa_eng/zxxx_662805/t1249761.shtml.

88. Ministry of Foreign Affairs, Sri Lanka, "Admiral Zheng He Promotes Tourism to Sri Lanka from the Port City of Taicang," January 23, 2013, https://www.mfa.gov.lk/admiral-zheng-he-promotes-tourism-to-sri-lanka-from-the-port-city-of-taicang/.
89. See Roncevert Ganan Almond, "Summits, Roads and Suspended Disbelief in Central Asia," *The Diplomat*, June 27, 2017, http://thediplomat.com/2017/06/summits-roads-and-suspended-disbelief-in-central-asia/.

CHAPTER FIVE

1. See Arjun Appadurai, "Introduction: Commodities and the Politics of Value," in *The Social Life of Things: Commodities in Cultural Perspective*, ed. Arjun Appadurai (Cambridge: Cambridge University Press, 1986), 3–63; and Igor Kopytoff, "The Cultural Biography of Things: Commoditization," in *The Social Life of Things: Commodities in Cultural Perspective*, ed. Arjun Appadurai (Cambridge: Cambridge University Press, 1986), 64–94
2. Hans Peter Hahn and Hadas Weiss, eds., *Mobility, Meaning and the Transformation of Things: Shifting Contexts of Material Culture through Time and Space* (Oxford, UK: Oxbow Books, 2013), loc. 269 of 5201, Kindle.
3. Hahn and Weiss, *Mobility, Meaning and the Transformation of Things*, loc. 269.
4. Michael Alram, "The History of the Silk Road as Reflected in Coins," *Parthica* 6 (2004): 47–68.
5. J. Keith Wilson and Michael Flecker, "Dating the Belitung Shipwreck," in *Shipwrecked: Tang Treasures and Monsoon Winds*, ed. Regina Krahl and Alison Effeny (Washington, DC: Arthur M. Sackler Gallery, Smithsonian Institution, and National Heritage Board, Singapore Tourism Board, 2010), 35–37.
6. For further information on this debate, see Regina Krahl and Alison Effeny, eds., *Shipwrecked: Tang Treasures and Monsoon Winds* (Washington, DC: Arthur M. Sackler Gallery, Smithsonian Institution, and National Heritage Board, Singapore Tourism Board, 2010).
7. Elizabeth Blair and James Delgado, interview with Linda Wertheimer, *Morning Edition*, NPR, May 4, 2011, http://www.npr.org/templates/transcript/transcript.php?storyId=135956044.
8. Kwa Chong-Guan, ed., *Early Southeast Asia Viewed from India: An Anthology of Articles from the "Journal of the Greater India Society"* (Delhi: Manohar, 2013).
9. For a fuller of account of the regional maritime trade in Southeast Asia, see John Miksic, *Singapore and the Silk Road of the Sea, 1300–1800* (Singapore: National University of Singapore Press, 2013).
10. Michael Flecker, "A 9th-Century Arab or Indian Shipwreck in Indonesian Waters," *International Journal of Nautical Archaeology* 29, no. 2 (2008): 199–217; and Sean Kingsley, "Editorial: Tang Treasures, Monsoon Winds

and a Storm in a Teacup," *The Undertow* (blog), March 13, 2011, https://wreckwatch.org/2011/03/13/editorial-tang-treasures-monsoon-winds-and-a-storm-in-a-teacup/.

11. Abdul Sheriff, "The Dhow Culture of the Western Indian Ocean," in *Journeys and Dwellings: Indian Ocean Themes in South Asia,* ed. Helene Basu (Hyderabad: Orient Blackswan, 2016), Kindle.
12. In respective order, see Himanshu Ray, "Seafaring in Peninsular India in the Ancient Period of Watercraft and Maritime Communities," in *Ships and the Development of Maritime Technology on the Indian Ocean,* ed. David Parkin and Ruth Barnes (London: Routledge, 2016), Kindle; Somasiri Devendra, "Pre-Modern Sri Lankan Ships," in *Ships and the Development of Maritime Technology on the Indian Ocean*; Susan Beckerleg, "Continuity and Adaptation by Contemporary Swahili Boatbuilders in Kenya," in *Ships and the Development of Maritime Technology on the Indian Ocean*; and Ruth Barnes, "Yami Boats and Boat Building in a Wider Perspective," in *Ships and the Development of Maritime Technology on the Indian Ocean.*
13. Sally Church, John Gebhardt, and Terry Little, "A Naval Architectural Analysis of the Plausibility of 450-Ft Treasure Ships," in *Zheng He and the Afro-Asian World,* ed. Chia Lin Sien and Sally Kathryn Church (Melaka: Perbadanan Muzium Melaka and International Zheng He Society, 2012), 15–47.
14. For further details of his acceptance speech, see President Sellapan Ramanathan Nathan, "Speech by HE President SR Nathan, President of Singapore, Welcoming Jewel of Muscat to Singapore" (speech, Singapore, July 3, 2010), http://jewelofmuscat.tv/speeches/speech-by-he-president-sr-nathan-president-of-singapore/.
15. Nathan, "Speech by HE President SR Nathan, President of Singapore."
16. "Celebrating 60 Years of Diplomatic Ties," *The Hindu*, March 24, 2016, http://www.thehindu.com/news/cities/Kochi/celebrating-60-years-of-diplomatic-ties/article7947457.ece.
17. John Miksic, introduction to *Ancient Silk Trade Routes: Selected Works from Symposium on Cross Cultural Exchanges and Their Legacies in Asia,* ed. Qin Dashu and Jian Yuan (Singapore: World Scientific, 2014), 1–17.
18. Michael Flecker, "The Thirteenth-Century Java Sea Wreck: A Chinese Cargo in an Indonesian ship," *Mariner's Mirror* 89, no. 4 (2003): 388–404.
19. Miksic, *Singapore and the Silk Road of the Sea, 1300–1800,* 91.
20. Miksic, *Singapore and the Silk Road of the Sea, 1300–1800,* 92.
21. Michael Walker, "Romano-Indian Rouletted Pottery in Indonesia," *Asian Perspectives: The Bulletin of the Far-Eastern Prehistory Association* 20, no. 2 (1980): 228–35.
22. Miksic lists the discovery at Ko Kho Khao and Laem Pho of five different types of glazed earthenware, including Basra turquoise, yellow-enameled luster white, and cobalt blue-and-white wares. For further details, see Miksic, *Singapore and the Silk Road of the Sea, 1300–1800,* 85.

23. For further details, see Robert Finlay, *The Pilgrim Art: Cultures of Porcelain in World History* (Berkeley: University of California Press, 2010), loc. 1717 of 6636, Kindle.
24. Michael Flecker, *The Archaeological Excavation of the 10th Century Intan Shipwreck* (Oxford, UK: Archaeopress, 2002).
25. Flecker, "Thirteenth-Century Java Sea Wreck," 388.
26. Flecker, "Thirteenth-Century Java Sea Wreck," 388–404.
27. Finlay, *Pilgrim Art*, loc. 145.
28. Finlay, *Pilgrim Art*, loc. 2819.
29. Finlay, *Pilgrim Art*, loc. 2805.
30. Stacey Pierson, *From Object to Concept: Global consumption and the Transformation of Ming Porcelain* (Hong Kong: Hong Kong University Press, 2014), 39.
31. Lisa Golombek and Eileen Reilly, "Safavid Society and the Ceramic Industry," in *Persian Pottery in the First Global Age: The Sixteenth and Seventeenth Centuries*, ed. Lisa Golombek, Robert Mason, Patricia Proctor, and Eileen Reilly (Leiden: Brill, 2014), 24.
32. William Bower Honey, *The Ceramic Art of China and Other Countries of the Far East* (London: Faber and Faber, 1945), 138.
33. Lisa Golombek, "Dominant Fashions and Distinctive Styles," in *Persian Pottery in the First Global Age: The Sixteenth and Seventeenth Centuries*, ed. Lisa Golombek, Robert Mason, Patricia Proctor, and Eileen Reilly (Leiden: Brill, 2014), 57–122; and Patricia Proctor, "The Measure of Faithfulness: The Chinese Models for Safavid Blue and White," in *Persian Pottery in the First Global Age*, 123–68.
34. Golombek, "Dominant Fashions and Distinctive Styles," 86.
35. Pierson, *From Object to Concept*, 44.
36. Tristan Mostert and Jan van Campen, *Silk Thread: China and the Netherlands from 1600* (Nijmegen, The Netherlands: Uitgeverij Vantilt, 2015), 38–39.
37. Mostert and van Campen, *Silk Thread*, 38–39.
38. Tijs Volker, *Porcelain and the Dutch East India Company: As Recorded in the Dagh-Registers of Batavia Castle, Those of Hirado and Deshima and Other Contemporary Papers, 1602–1682* (Leiden: Brill, 1954).
39. Finlay, in *Pilgrim Art*, offers *kendi* and pilgrim flasks as a further example, wherein form and function changed over time as they migrated across great distances and through different religious and cultural contexts. Carried by Japanese pilgrims and sold by Javanese merchants, such flasks decorated the mosques of the Islamic world, the dining tables of Ottoman courts, and the houses of Portuguese Christians living in Goa.
40. Finlay, *Pilgrim Art*, loc. 178.
41. See, e.g., Lily Kong and Justin O'Connor, *Creative Economies, Creative Cities: Asian-European Perspectives* (New York: Springer, 2009).
42. Wilfried Wang, *Culture: City* (Baden, Germany: Lars Müller, 2013).

43. Christine Sylvester, *Art/Museums: International Relations Where We Least Expect It* (Boulder, CO: Paradigm Publishers, 2009); and Gail Lord and Ngaire Blankenberg, eds., *Cities, Museums and Soft Power* (Washington, DC: American Alliance of Museums, 2015).
44. Gail Lord and Ngaire Blankenberg, "Introduction: Why Cities, Museums and Soft Power," in *Cities, Museums and Soft Power*, ed. Gail Lord and Ngaire Blankenberg (Washington, DC: American Alliance of Museums, 2015), 20–24.
45. Jeff Adams, "The Role of Underwater Archaeology in Framing and Facilitating the Chinese National Strategic Agenda," in *Cultural Heritage Politics in China*, ed. Tami Blumenfield and Helaine Silverman (New York: Springer, 2013), 261–82. See also Andrew Erickson and Kevin Bond, "Archaeology and the South China Sea," *The Diplomat*, July 20, 2015, http://thediplomat.com/2015/07/archaeology-and-the-south-china-sea/.
46. Sun Ye and Karl Wilson, "Oceans of Heritage," *ChinaDaily Asia*, April 24, 2015, http://www.chinadailyasia.com/asiaweekly/2015–04/24/content_15257382.html, quoting Jiang Bo, director of the Institute of Underwater Archaeology at the National Conservation Center for Underwater Cultural Heritage.
47. Sun Ye and Wilson, "Oceans of Heritage."
48. For further details, see "China Maritime Silk Road Museum," *China Heritage Newsletter: China Heritage Project* (Australian National University), March 2005, http://www.chinaheritagequarterly.org/articles.php?searchterm=001_maritimesilk.inc&issue=001.
49. For further details, see Chinese Academy of Social Sciences, "Maritime Silk Road on Show in Fujian," Chinese Archaeology: Institute of Archaeology, http://www.kaogu.net.cn/en/News/Academic_activities/2014/1212/48579.html.
50. *ChinaDaily*, "Quanzhou to Build Maritime Silk Road Museum," January 5, 2016, http://www.chinadaily.com.cn/m/fujian/jinjiang/2016–01/05/content_22959273.htm.
51. *I Cross China*, "'Maritime Silk Road' Themed Subway Hits Ningbo," May 21, 2015, http://www.icrosschina.com/sizzling/2015/0521/12875.shtml.
52. For further details, see *China.org.cn*, "Xi Tours 'Living Fossil of Silk Road' Bukhara in Uzbekistan," June 22, 2016, http://www.china.org.cn/world/2016–06/22/content_38716368.htm.
53. The broader aims are recorded in the following public statement made by the China National Silk Museum:

> Luxor and Zhejiang are similar in the sense that they were both places of significant cultural value to their countries. Luxor, the ancient Egyptian city, known for its heritage that trace all the way back to the Ancient Egyptian capital of Thebes, contains many important ancient Egyptian artifacts. Zhejiang is the 'Silk Capital.' The 7000 year old

Hemudu Civilization, the 5000 year old Liangzhu Civilization, and the 4000 year old Qianshangyan Civilization were periods in which silk weaving was established and improved, which paved way for Zhejiang to be the undisputed center of silk trade on both the land and maritime Silk Roads. The prevalence of silk around the world advanced the economic and cultural ties between East and West. The purpose of the exhibition is as thus: to enable locals to glimpse at China's 5000 year old silk culture, and to promote the cultural exchange and collaboration between Zhejiang and Luxor.

Available at China National Silk Museum, "'Silk on the Silk Road' Exhibition Unveils in Luxor," March 26, 2016, http://en.chinasilkmuseum.com/news/detail_605.html. See also China National Silk Museum, "'Silk on the Silk Road' Exhibition Unveils in Cairo," March 26, 2016, http://en.chinasilkmuseum.com/news/detail_604.html.

54. The public statement for the exhibition's opening ceremony stated that "Qatar, a country situated on the intersection of Asia, Africa and Europe, is both an important Western Asian nation and an important town in the Western segment of the Silk Road. Under the 'One generation, One road' banner proposed by President Xi Jinping of China, China and Qatar, countries situated at either ends of the ancient silk road, began to accelerate trade and had recently decided to host the 2016 China-Qatar Cultural Year, an event intended to advance cultural exchange between these two nations." See China National Silk Museum, "'Silk on the Silk Road—Chinese Silk Exhibition' Unveils in Qatar," March 23, 2016, http://en.chinasilkmuseum.com/news/detail_603.html.
55. Mark Vlasic and Helga Turku, "'Blood Antiquities,'" *Journal of International Criminal Justice* 14, no. 5 (2016): 1045.
56. Amr Al Azm, "The Pillaging of Syria's Cultural Heritage," Middle East Institute, May 22, 2015, http://www.mei.edu/content/at/pillaging-syrias-cultural-heritage.
57. Joe Parkinson, Ayla Albayrak, and Duncan Mavin, "Syrian 'Monuments Men' Race to Protect Antiquities as Looting Bankrolls Terror," *Wall Street Journal*, February 10, 2015, http://www.wsj.com/articles/syrian-monuments-men-race-to-protect-antiquities-as-looting-bankrolls-terror-1423615241.
58. See, e.g., Peter Stone, Joanne Farchakh Bajjaly, and Robert Fisk, *The Destruction of Cultural Heritage in Iraq* (Woodbridge, UK: Boydell Press, 2008); France Desmarais, ed., *Countering Illicit Traffic in Cultural Goods: The Global Challenge of Protecting the World's Heritage* (Paris: International Council of Museums International Observatory on Illicit Traffic in Cultural Goods, 2015); and Vlasic and Turku, "'Blood Antiquities.'"
59. In a different publication genre, equally revealing and intriguing were the exploits of a former CIA officer-turned-curator and his desire to acquire an ancient statue of Athena, as portrayed by Arthur Houghton in his 2016

novel *Dark Athena* (n.p.: CreateSpace Independent Publishing Platform, 2016).

60. Oliver Moody, "Isis Fills War Chest by Selling Looted Antiquities to the West," *The Times* (London), December 17, 2014, http://www.thetimes.co.uk/tto/news/world/middleeast/article4299572.ece; and Simon Cox, "The Men Who Smuggle the Loot That Funds IS," *BBC News*, February 17, 2015, http://www.bbc.com/news/magazine-31485439.
61. See Vlasic and Turku, "'Blood Antiquities'"; Martin Chulov, "How an Arrest in Iraq Revealed Isis's $2bn Jihadist Network," *The Guardian*, June 16, 2014, https://www.theguardian.com/world/2014/jun/15/iraq-isis-arrest-jihadists-wealth-power; and Duncan Mavin, "Calculating the Revenue from Antiquities to Islamic State," *Wall Street Journal*, February 11, 2015, http://www.wsj.com/articles/calculating-the-revenue-from-antiquities-to-islamic-state-1423657578.
62. Donna Yates, Simon Mackenzie, and Emiline Smith, "The Cultural Capitalists: Notes on the Ongoing Reconfiguration of Trafficking Culture in Asia," *Crime Media Culture* 13, no. 2 (2017): 245–54.
63. "Mad about Museums," *The Economist*, January 6, 2014, http://www.economist.com/news/special-report/21591710-china-building-thousands-new-museums-how-will-it-fill-them-mad-about-museums; and Dan Howarth, "China 'Can't Buy Culture' with Museum Boom, Say Critics," *Dezeen*, December 11, 2015, http://www.dezeen.com/2015/12/11/new-chinese-museums-construction-boom-opening-money-cant-buy-culture-china/.
64. Frank Langfitt, "China Builds Museums, but Filling Them Is Another Story," *All Things Considered*, May 21, 2013, http://www.npr.org/sections/parallels/2013/05/21/185776432/china-builds-museums-but-will-the-visitors-come.
65. Michael St. Clair, *The Great Chinese Art Transfer: How So Much of China's Art Came to America* (Madison, NJ: Fairleigh Dickinson University Press, 2016), 18.
66. For further details, see St. Clair, *Great Chinese Art Transfer*; and Paola Demattè, "Emperors and Scholars: Collecting Culture and Late Imperial Antiquarianism," in *Collecting China: The World, China, and a Short History of Collecting*, ed. Vimalin Rujivacharakul (Newark: University of Delaware Press, 2011), 165–75.
67. Steven Gallagher, "'Purchased in Hong Kong': Is Hong Kong the Best Place to Buy Stolen or Looted Antiquities?," *International Journal of Cultural Property* 24 (2017): 483.
68. Brent Beardsley, Jorge Becerra, Federico Burgoni, Bruce Holley, Daniel Kessler, Federico Muxi, Matthias Naumann, Tjun Tang, André Xavier, and Anna Zakrzewski, "Global Wealth 2015: Winning the Growth Game," https://www.bcgperspectives.com/content/articles/financial-institutions-growth-global-wealth-2015-winning-the-growth-game/?chapter=2#chapter2_section2.

69. Pierre Bourdieu, *The Field of Cultural Production: Essays on Art and Literature* (New York: Columbia University Press, 1993).
70. Denis Byrne, "The Problem with Looting: An Alternative Perspective on Antiquities Trafficking in Southeast Asia," *Journal of Field Archaeology* 41, no. 3 (2016): 347.
71. UNIDROIT is the International Institute for the Unification of Private Law. See International Council of Museums, "Fighting Illicit Traffic," accessed July 30, 2016, http://icom.museum/programmes/fighting-illicit-traffic/.

CHAPTER SIX

1. See Fernand Braudel, *The Structure of Everyday Life*, vol. 1 of *Civilization and Capitalism: 15th–18th Century* (Berkeley: University of California Press, 1992); Braudel, *The Mediterranean and the Mediterranean World in the Age of Philip II*, vol. 1 (Berkeley: University of California Press, 2012); Immanuel Wallerstein, *The Modern World-System I: Capitalist Agriculture and the Origins of the European World-Economy in the Sixteenth Century* (Berkeley: University of California Press, 2011); Immanuel Wallerstein, *The Modern World System II: Mercantilism and the Consolidation of the European World-Economy, 1600–1750* (Berkeley: University of California Press, 2011); and Immanuel Wallerstein, *The Modern World-System III: The Second Era of Great Expansion of the Capitalist World-Economy, 1730s–1840s* (New York: Academy Press, 2011).
2. See, e.g., Walter Mignolo, *The Darker Side of the Renaissance: Literacy, Territoriality, and Colonization* (Ann Arbor: University of Michigan Press, 2003).
3. Andre Gunder Frank, *ReOrient: Global Economy in the Asian Age* (Berkeley: University of California, 1998), Kindle; Michael Pearson, *The Indian Ocean* (London: Routledge, 2010); and Philippe Beaujard, "The Worlds of the Indian Ocean," in *Trade, Circulation, and Flow in the Indian Ocean World*, ed. Michael Pearson (Houndmills, UK: Palgrave Macmillan, 2015), loc. 515 of 4634, Kindle.
4. Frank, *ReOrient: Global Economy in the Asian Age*, loc. 523 of 9411.
5. Pearson, *Indian Ocean*, 8.
6. Pearson, *Indian Ocean*, 16.
7. Christopher Beckwith, *Empires of the Silk Road: A History of Central Eurasia from the Bronze Age to the Present* (Princeton, NJ: Princeton University Press, 2009), loc. 2071 of 11575, Kindle.
8. For more details, see Philippe Beaujard, *L'Océan Indien, au cœur des globalisations de l'Ancien Monde (7e–15e siècles)* (Paris: Collin, 2012): and Beaujard, "Worlds of the Indian Ocean."
9. Sing Chew, "The Southeast Asian Connection in the First Eurasian World Economy, 200BCE–CE 500," in *Trade, Circulation, and Flow in the Indian Ocean*, ed. Michael Pearson (Houndmills, UK: Palgrave Macmillan, 2015), loc. 645 of 4634, Kindle.

10. Beckwith, *Empires of the Silk Road*, loc. 6206.
11. Xinru Liu, *The Silk Road in World History* (Oxford: Oxford University Press, 2010), loc. 1119 of 3296, Kindle.
12. Ross Dunn, *The Adventures of Ibn Battuta* (Berkeley: University of California Press, 1986), quoted in Michael Pearson, *The Indian Ocean* (London: Routledge, 2010), 77.
13. See John Miksic, *Singapore and the Silk Road of the Sea, 1300–1800* (Singapore: National University of Singapore Press, 2013).
14. Janice Stargardt, "Indian Ocean Trade in the Ninth and Tenth Centuries: Demand, Distance, and Profit," *South Asian Studies* 30, no. 1 (2014): 35–55.
15. For further details, see Geoff Wade, "Chinese Engagement with the Indian Ocean during the Song, Yuan and Ming Dynasties (Tenth to Sixteenth Centuries)," in *Trade, Circulation, and Flow in the Indian Ocean World*, ed. Michael Pearson (Houndmills, UK: Palgrave Macmillan, 2015), loc. 1227 of 4634, Kindle.
16. See Tansen Sen, "Diplomacy, Trade and the Quest for the Buddha's Tooth: The Yongle Emperor and Ming China's South Asian Frontier," in *Ming China: Courts and Contacts, 1400–1450*, ed. Jessica Harrison-Hall, Craig Clunas, and Luk Yu-ping (London: British Museum, 2016), 26–36.
17. See Anthony Reid, *Southeast Asia in the Age of Commerce, 1450–1680*, vol. 2 (New Haven, CT: Yale University Press, 1993). As Fernando illustrates, this decline in Melaka's significance would reverse some 160 years later as trade with India resumed and grew quickly from 1620 onward. See M. Fernando, "Continuity and Change in Maritime Trade in the Straits of Melaka in the Seventeenth and Eighteenth Centuries," in *Trade, Circulation, and Flow in the Indian Ocean World*, ed. Michael Pearson (Houndmills, UK: Palgrave Macmillan, 2015), Kindle.
18. Tansen Sen, "The Impact of Zheng He's Expeditions on Indian Ocean Interactions," *Bulletin of the School of Oriental and African Studies* 79, no. 3 (2016): 621.
19. See Christopher Bayly, *The Birth of the Modern World 1780–1914: Global Connections and Comparisons* (Malden, MA: Blackwell, 2012); Takeshi Hamashita, *China, East Asia and the Global Economy: Regional and Historical Perspectives* (London: Routledge, 2008); and Kenneth Pomeranz, *The Great Divergence: China, Europe and the Making of the Modern World Economy* (Princeton, NJ: Princeton University Press, 2001).
20. S. Frederick Starr, *Lost Enlightenment: Central Asia's Golden Age from the Arab Conquest to Tamerlane* (Princeton, NJ: Princeton University Press, 2013), locs. 1072–1129 of 15916, Kindle.
21. Starr, *Lost Enlightenment*, loc. 1133, 1146.
22. See Starr, *Lost Enlightenment*, loc. 872.
23. Beckwith, *Empires of the Silk Road*, loc. 5545.
24. Beckwith, *Empires of the Silk Road*, loc. 6099.

25. This is presented as an overarching trend, and a number of examples can be cited that lie outside such framings, such as Xi'an.
26. Benedict Anderson, *Imagined Communities: Reflections on the Origin and Spread of Nationalism* (Ithaca, NY: Cornell University Press, 1991). See also Penny Edwards, *Cambodge: The Cultivation of a Nation, 1860–1945* (Honolulu: University of Hawai'i Press, 2007); Anne Blackburn, *Locations of Buddhism: Colonialism and Modernity in Sri Lanka* (Chicago: University of Chicago Press, 2010); and Tapati Guha-Thakurta, *Monuments, Objects, Histories: Institutions of Art in Colonial and Postcolonial India* (New York: Columbia University Press, 2004).
27. The historic center of Bukhara was inscribed in 1993 on the basis of three criteria for outstanding universal value: "Criterion (ii): The example of Bukhara in terms of its urban layout and buildings had a profound influence on the evolution and planning of towns in a wide region of Central Asia; Criterion (iv): Bukhara is the most complete and unspoiled example of a medieval Central Asian town which has preserved its urban fabric to the present day; Criterion (vi): Between the 9th and 16th centuries, Bukhara was the largest centre for Muslim theology, particularly on Sufism, in the Near East, with over two hundred mosques and more than a hundred madrasahs." See UNESCO, "Historic Centre of Bukhara," http://whc.unesco.org/en/list/602.
28. Inscribed in 2003, the Mausoleum of Khoja Ahmed Yasawi was listed on the basis of the following statements of significance: "Criterion (i): The Mausoleum of Khoja Ahmed Yasawi is an outstanding achievement in the Timurid architecture, and it has significantly contributed to the development of Islamic religious architecture; Criterion (iii): The mausoleum and its property represent an exceptional testimony to the culture of the Central Asian region, and to the development of building technology; Criterion (iv): The Mausoleum of Khoja Ahmed Yasawi was a prototype for the development of a major building type in the Timurid period, becoming a significant reference in the history of Timurid architecture." See UNESCO, "Mausoleum of Khoja Ahmed Yasawi," http://whc.unesco.org/en/list/1103.
29. Kwa Chong-Guan, ed., *Early Southeast Asia Viewed from India: An Anthology of Articles from the "Journal of the Greater India Society"* (Delhi: Manohar, 2013). Christopher Beckwith also explains the factors that led to the emergence of the littoral system in Asia at the time when trade networks along the overland Silk Road were declining. See Beckwith, *Empires of the Silk Road.*
30. See, e.g., V. Bulkin, Leo Klejn, and G. S. Lebedev, "Attainments and Problems of Soviet Archaeology," *World Archaeology* 13, no. 3 (1982): 272–95.
31. See Ali Mozaffari, *Forming National Identity in Iran: The Idea of Homeland Derived from Ancient Persia and Islamic Imaginations of Peace* (London: I. B. Tauris, 2014), Kindle.

32. Minoo Moallem, *Persian Carpets: The Nation as a Transnational Commodity* (Oxford: Routledge, 2018), p. 3, Kindle; and Hamid Dabashi, *Iran without Borders: Towards a Critique of the Postcolonial Nation* (London: Verso, 2016).
33. See, e.g., Tansen Sen, "Silk Road Diplomacy—Twists, Turns and Distorted History," YaleGlobal Online, September 23, 2014, http://yaleglobal.yale.edu/content/silk-road-diplomacy-twists-turns-and-distorted-history.
34. Hvistendahl also details the military nature of Zheng He's voyages. See Mara Hvistendahl, "Rebuilding a Treasure Ship," *Archaeology* 61, no. 2 (2008): 40–45.
35. Geoff Wade, "The Zheng He Voyages: A Reassessment" (ARI Working Paper No. 31, Asia Research Institute, October 2004), 14.
36. Sen, "Diplomacy, Trade and the Quest for the Buddha's Tooth," 30; Wade, "Zheng He Voyages," 16; and Wade, "Chinese Engagement with the Indian Ocean," loc. 1443.
37. Geoff Wade has also argued that the violent advancement of Genghis Khan from the north would lead to an increase in maritime infrastructure and trade: "The expansion of the Mongols across Eurasia under Genghis Khan from the beginning of the thirteenth century was to have profound effects on the Chinese polities. Gaining control of China allowed the Mongol Yuan administration to set their sights even wider and, given that the sea lay between them and other potential territories, the importance of naval forces became even greater in the last two decades of the thirteenth century. The relevance of the Yuan naval expeditions in this period to world history and their effects, particularly on maritime Asia, have been greatly understudied." See Wade, "Chinese Engagement with the Indian Ocean," loc. 1286.
38. Michael Pearson, "Introduction: Maritime History and the Indian Ocean World," in *Trade, Circulation, and Flow in the Indian Ocean World*, ed. Michael Pearson (Houndmills, UK: Palgrave Macmillan, 2015), locs. 154, 160 of 4634, Kindle.
39. Thomas Vernet, "East African Travelers and Traders in the Indian Ocean: Swahili Ships, Swahili Mobilities, ca. 1500–1800," in *Trade, Circulation, and Flow in the Indian Ocean World*, ed. Michael Pearson (Houndmills, UK: Palgrave Macmillan, 2015), Kindle.
40. Silk Road Universities Network, "Constitution," http://www.sun-silkroadia.org/eng/info/03.php.
41. Shannon Tiezzi provides a nice overview of this issue in the article "China's Academic Battle for the South China Sea," *The Diplomat*, February 25, 2014, http://thediplomat.com/2014/02/chinas-academic-battle-for-the-south-china-sea/.
42. For a detailed account of the links between underwater archaeology and marine research, see Jeff Adams, "The Role of Underwater Archaeology in Framing and Facilitating the Chinese National Strategic Agenda," in

Cultural Heritage Politics in China, ed. Tami Blumenfield and Helaine Silverman (New York: Springer, 2013), 261–82.

43. Toshi Yoshihara, "China's 'Soft' Naval Power in the Indian Ocean," *Pacific Focus* 25, no. 1 (2010): 59–88; and James Holmes and Toshi Yoshihara, "Soft Power Goes to Sea," *American Interest* (Spring 2008): 66–67.
44. This was illustrated in the press release issued for the center's launch ceremony:

    > All the experts believed that the construction of 'One Road and One Belt' was the Chinese government's major initiative to dispel the misgiving from other countries. It also broke the pattern in which a strong country was bound to pursue hegemony, manifesting the connectivity between China Dream and the World Dream. The Academic Belt of the Silk Road was a response to the national strategy for Chinese great rejuvenation. It focused on integrating academic research related to the Silk Road, improving the exchanges between experts from different countries and their ability of knowledge production. Tackling key problems by a collaborative innovation which was transnational, transregional and transboundary, it aimed at providing the foresight and intellectual support for the Road and Belt Initiative.

    See Xi'an Jiaotong University, "Collaborative Innovation Centre of Silk Road Economic Belt Research Started by XJTU," January 23, 2015, http://en.xjtu.edu.cn/info/1044/1572.htm.
45. "Collaborative Innovation Centre of Silk Road Economic Belt Research Started by XJTU."
46. Ding Yi, "Huaqiao University—A Think Tank for 'One Belt, One Road' Strategy," *Sino-US.com*, June 10, 2015, http://www.sino-us.com/433/15021790078.html.
47. In January 2017, for example, a delegation from Timor-Leste visited the university to discuss future cooperation in cultural and academic areas. See City University of Macau, "Macau 'One Belt, One Road' Research Center," May 2017, http://www.cityu.edu.mo/moborrc/en/home.
48. Within the City University of Hong Kong Research Centre on One Belt One Road, for example, a program of building "cultural intelligence" has been established to effectively deal with cross-cultural settings. See City University of Hong Kong—Research Centre on One-Belt-One-Road, "Introduction," May 2017, http://www.cb.cityu.edu.hk/obor/.
49. *Xinhua News*, "Cambodia Launches China-Backed Maritime Silk Road Research Centre," June 13, 2016, http://news.xinhuanet.com/english/2016-06/13/c_135432581.htm.
50. See *Hanban News*, "The First 'Confucius Institute of Maritime Silk Road' Established in Thailand," July 6, 2015, http://english.hanban.org/article/2015-07/06/content_608534.htm.

51. Full text available at Silk Road Universities Network, "Purpose and Background," http://www.sun-silkroadia.org/eng/info/02.php.
52. The national framing undertaken by scholars employed in institutions across the region is reflected in the published literature and was evident in the 2014 Asia-Pacific Regional Conference on Underwater Cultural Heritage in Hawai'i. For program details, see National Marine Sanctuary Foundation and University of Hawai'i, "Final APCONF 2014 Program," Asia-Pacific Conference on Underwater Cultural Heritage, December 2, 2016, http://www.apconf.org/wp-content/uploads/Final-APCONF-2014-program.pdf.
53. See *News.gov.hk*, "Treasure Trade: Maritime Silk Road," November 20, 2016, http://www.news.gov.hk/en/city_life/html/2016/11/20161117_154216.shtml; and Government of the Hong Kong Special Administrative Region, "Museum of History Exhibition Reveals Development of China's Maritime Silk Road (with Photos)," press release, October 25, 2016, http://www.info.gov.hk/gia/general/201610/25/P2016102500265.htm.
54. See Maritime Silk Road Society, "Leaders from Top-Notch Enterprises Speak at 'The Belt and Road Initiative—Combining Hard Power with Soft Power' Conference," February 27, 2016, http://maritimesilkroad.org.hk/en/news/newsReleasesDetail/102.
55. Since 2014 the National University of Singapore has run the Muhammad Alagil Arabia Asia Conference series, including Arabia-Asia Relations, Then and Now (2014); Silk Roads, Muslim Passages: The Islam Question (2015); and China-Arabia Encounters and Engagements (2016).
56. Lim Tai-Wei, Henry Chan Hing Lee, Katherine Tseng Hui-Yi, and Lim Wen Xi, eds., *China's One Belt One Road Initiative* (London: Imperial College Press, 2016).
57. Lim Tai-Wei, introduction to *China's One Belt One Road Initiative*, ed. Lim Tai-Wei, Henry Chan Hing Lee, Katherine Tseng Hui-Yi, and Lim Wen Xi (London: Imperial College Press, 2016), 8.
58. Katherine Tseng Hui-Yi, "The South China Sea and the Maritime Silk Road Proposal: Conflicts Can Be Transformed," in *China's One Belt One Road Initiative*, ed. Lim Tai-Wei, Henry Chan Hing Lee, Katherine Tseng Hui-Yi, and Lim Wen Xi (London: Imperial College Press, 2016), 141.
59. Lim Tai-Wei, introduction to *China's One Belt One Road Initiative*, 5.
60. Tan Ta Sen, "Cheng Ho Spirit and World Dream," in *China's One Belt One Road Initiative*, ed. Lim Tai-Wei, Henry Chan Hing Lee, Katherine Tseng Hui-Yi, and Lim Wen Xin (London: Imperial College Press, 2016), 57–59.
61. A Google Scholar search conducted in February 2017 for articles with "One Belt One Road" in the title identified the following totals: 22 in 2014; 547 in 2015; and 1,030 in 2016. Articles with titles including "The Silk Road Economic Belt" totaled 620 in 2014; 1,520 in 2015; and 1,550 in 2016.

CHAPTER SEVEN

1. *The Economist*, "A Russian Orchestra Plays Bach and Prokofiev in the Ruins of Palmyra," May 6, 2016, https://www.economist.com/europe/2016/05/06/a-russian-orchestra-plays-bach-and-prokofiev-in-the-ruins-of-palmyra.
2. Justin Carissimo, "Isis Propaganda Video Shows 25 Syrian Soldiers Executed by Teenage Militants in Palmyra," *The Independent*, July 5, 2015, https://www.independent.co.uk/news/world/middle-east/isis-propaganda-video-shows-20-syrian-government-soldiers-executed-in-palmyra-by-young-islamists-10366533.html.
3. *The Economist*, "Russian Orchestra."
4. Immanuel Wallerstein, *Geopolitics and Geoculture: Essays on the Changing World-System* (Cambridge: Cambridge University Press, 1991).
5. Ulf Hannerz, "Geocultural Scenarios," in *Frontiers of Sociology*, ed. Peter Hedstrom and Bjorn Wittrock (Leiden: Brill, 2009), 268.
6. Hannerz, "Geocultural Scenarios," 268.
7. David Shambaugh and Zhang Weiwei are among those who have argued that we need to understand China's rise by conceptualizing it as a civilizational state. See David Shambaugh, *China Goes Global: The Partial Power* (Oxford: Oxford University Press, 2013); and Zhang Weiwei, *The China Wave: The Rise of a Civilisational State* (Hackensack, NJ: World Century, 2012), 2.
8. Randall Collins, "Civilizations as Zones of Prestige and Social Contact," *International Sociology* 16, no. 3 (2001): 422.
9. Alfred Rieber, *The Struggle for the Eurasian Borderlands: From the Rise of Early Modern Empires to the End of the First World War* (Cambridge: Cambridge University Press, 2014), loc. 231 of 18946, Kindle.
10. Rieber, *Struggle for the Eurasian Borderlands*, loc. 221.
11. Bruce Trigger, *A History of Archaeological Thought* (Cambridge: Cambridge University Press, 1989).
12. William Callahan, "China's 'Asia Dream': The Belt Road Initiative and the New Regional Order," *Asian Journal of Comparative Politics* 1, no. 3 (2016): 226.
13. Antara Ghosal Singh, "China's Soft Power Projection across the Oceans," *Maritime Affairs: Journal of the National Maritime Foundation of India* 12, no. 1 (2016): 25–37; and Timo Kivimäki, "Soft Power and Global Governance with Chinese Characteristics," *Chinese Journal of International Politics* 7, no. 4 (2014): 421–77.
14. Kivimäki, "Soft Power and Global Governance with Chinese Characteristics," 421–77.
15. William Callahan, *China: The Pessoptimist Nation* (Oxford: Oxford University Press, 2010), loc. 392 of 5783, Kindle.
16. Xi Jinping, "Speech by H.E. Xi Jinping President of the People's Republic of China at UNESCO Headquarters" (speech, Paris, March 27, 2014), http://rs.chineseembassy.org/eng/xwdt/t1142560.htm.

17. Ministry of Foreign Affairs of the People's Republic of China, "Xi Jinping Delivers Important Speech at Headquarters of the League of Arab States, Stressing to Jointly Create a Bright Future for Development of China-Arab Relations and Promote National Rejuvenation of China and Arab States to Form More Convergence," January 22, 2016, http://www.fmprc.gov.cn/mfa_eng/topics_665678/xjpdstajyljxgsfw/t1334587.shtml.
18. See Xi Jinping, "New Blueprint for Global Economic Growth"; and Xi Jinping, "President Xi's Speech to Davos in Full" (speech, Davos, January 17, 2017), https://www.weforum.org/agenda/2017/01/full-text-of-xi-jinping-keynote-at-the-world-economic-forum.
19. This spirit or resilience and renaissance builds on the political discourse from the late 1990s onward. Note, for example, the following excerpt from a speech by Hu Jintao given in Pretoria, South Africa, on February 7, 2007:

> Six hundred years ago, Zheng He, a famed Chinese navigator of the Ming Dynasty, headed a large convoy which sailed across the ocean and reached the east coast of Africa four times. They brought to the African people a message of peace and goodwill, not swords, guns, plunder or slavery. For more than one hundred years in China's modern history, the Chinese people were subjected to colonial aggression and oppression by foreign powers and went through similar suffering and agony that the majority of African countries endured. From the mid-19th century to the mid-20th century, the Chinese people launched a heroic struggle to fight colonial aggression and foreign oppression, achieve independence and liberation and build a new China of the Chinese people. Having realized this century-long historic mission, the Chinese people today are working as one to make life better for themselves. Because of the sufferings they experienced and the struggle they launched, something they will never forget, the Chinese people are most strongly opposed to colonialism, oppression, and slavery of all manifestations. Because of this, the Chinese people have the most profound sympathy for all other nations in their pursuit of independence, happiness and their aspirations.

See Hu Jintao, "Enhance China-Africa Unity and Cooperation to Build a Harmonious World" (speech, Pretoria, February 7, 2007), http://www.fmprc.gov.cn/mfa_eng/wjdt_665385/zyjh_665391/t298174.shtml.
20. Zhao Tingyang, *The Tianxia System: An Introduction to the Philosophy of a World Institution* (Nanjing: Jiangsu Jiaoyu Chubanshe, 2005) (published in Chinese); Yan Xuetong, *Ancient Chinese Thought, Modern Chinese Power*, ed. Daniel Bell and Sun Zhe (Princeton, NJ: Princeton University Press, 2011), Kindle; and Howard French, *Everything under the Heavens: How the Past Helps Shape China's Push for Global Power* (New York: Knopf, 2017), Kindle.
21. Ban Wang, "The Moral Vision in Kong Youwei's Book of the Great Community," in *Chinese Visions of World Order: Tianxia, Culture, and World*

Politics, ed. Ban Wang (Durham, NC: Duke University Press, 2017), pp. 87–105, Kindle.

22. Ban Wang, "Moral Vision in Kong Youwei's Book of the Great Community"; and Prasenjit Duara, "The Chinese World Order and Planetary Sustainability," in *Chinese Visions of World Order: Tianxia, Culture and World Politics*, ed. Ban Wang (Durham, NC: Duke University Press, 2017), pp. 65–83, Kindle.
23. French, *Everything under the Heavens*, 2.
24. Geoff Wade, "Civilisational Rhetorical and the Obfuscation of Power Politics," in *Sacred Mandates: Asian International Relations since Chinggis Khan*, ed. Timothy Brook, Michael van Walt can Praag, and Miek Boltjes (Chicago: University of Chicago Press, 2018), p. 76, Kindle.
25. Edward Vickers, "A Civilizing Mission with Chinese Characteristics? Education, Colonialism and Chinese State Formation in Comparative Perspective," in *Constructing Modern Asian Citizenship*, ed. Edward Vickers and Krishna Kumar (Oxford: Routledge, 2015), 50–79; and Stevan Harrell, ed., *Cultural Encounters on China's Ethnic Frontiers* (Seattle: University of Washington, 1995), Kindle.
26. Zhang Weiwei, *The China Horizon: Glory and Dream of Civilisational State* (Hackensack, NJ: World Century Publishing, 2016), pp. 157–58, Kindle.
27. Timothy Brook, Michael van Walt van Praag, and Miek Boltjes, eds., *Sacred Mandates: Asian International Relations since Chinggis Khan* (Chicago: University of Chicago Press, 2018), p. 5, Kindle.
28. Daniel Bell, "Realising Tianxia: Traditional Values and China's Foreign Policy," in *Chinese Visions of World Order: Tianxia, Culture and World Politics*, ed. Ban Wang (Durham, NC: Duke University Press, 2017), pp. 129–46, Kindle; and Yan Xuetong, *Ancient Chinese Thought, Modern Chinese Power.*
29. Alister Miskimmon, Ben O'Loughlin, and Laura Roselle, eds., *Strategic Narratives: Communication Power and the New World Order* (London: Routledge, 2013), locs. 217–21 of 551, Kindle.
30. Ravi Bhoothalingam, "The Silk Road as a Global Brand," *China Report* 52, no. 1 (2016): 52.
31. See also Michael Share, "The Great Game Revisited: Three Empires Collide in Chinese Turkestan," *Europe-Asia Studies* 67, no. 7 (2015): 1102–29.
32. Andrew Korybko, "What Could India's 'Cotton Route' Look Like?," *Sputnik International*, March 29, 2015, http://sputniknews.com/columnists/20150329/1020170021.html.
33. As Plets explains, immediately after the event, various programs of restoration and conservation were proposed by Russian cultural institutions, notably the State Hermitage Museum. See Gertjan Plets, "Violins and Trowels for Palmyra: Post-Conflict Heritage Politics," *Anthropology Today* 33, no. 4 (2017): 18–22.
34. Hannerz, "Geocultural Scenarios," 276.

35. For a more detailed account of India's regional ambitions, see Teresita Schaffer and Howard Schaffer, *India at the Global High Table: The Quest for Regional Primacy and Strategic Autonomy* (Uttar Pradesh, India: Harper Collins, 2016).
36. Anna Tsing, *Friction: An Ethnography of Global Connection* (Princeton, NJ: Princeton University Press, 2005), loc. 229 of 7725, Kindle.
37. Tsing, *Friction*, loc. 261.
38. James Millward, *Eurasian Crossroads: A History of Xinjiang* (London: Hurst and Co., 2007), 357.
39. See, e.g., Gertjan Plets, "Ethno-nationalism, Asymmetric Federalism and Soviet Perceptions of the Past: (World) Heritage Activism in the Russian Federation," *Journal of Social Archaeology* 15, no. 1 (2015): 67–93; and Gertjan Plets, "Heritage Bureaucracies and the Modern Nation State: Towards an Ethnography of Archaeological Systems of Government," *Archaeological Dialogues* 23, no. 2 (2016): 193–213.
40. Parag Khanna, *Connectography: Mapping the Global Network Revolution* (London: Weidenfeld and Nicolson, 2016).
41. James Millward, *The Silk Road: A Very Short Introduction* (Oxford: Oxford University Press, 2013), p. 117, Kindle.

Bibliography

Abu-Lughod, Janet. *Before European Hegemony: The World System A.D. 1250–1350.* New York: Oxford University Press, 1991.

Adams, Jeff. "The Role of Underwater Archaeology in Framing and Facilitating the Chinese National Strategic Agenda." In *Cultural Heritage Politics in China*, edited by Tami Blumenfield and Helaine Silverman, 261–82. New York: Springer, 2013.

AidData. "CRBC Builds Railway Link from Mombasa to Nairobi—Concessional Loan (Linked to Project ID #31777)." http://china.aiddata.org/projects/35232?iframe=y.

———. "Geospatial Dashboard." http://china.aiddata.org/geospatial_dashboardq=Cambodia&1=12.706042123782732, 104.23112999999998,9.

Akagawa, Natsuko. *Heritage Conservation and Japan's Cultural Diplomacy: Heritage, National Identity and National Interest.* London: Routledge, 2014.

Akbari, Suzanne. "Introduction: East, West, and In-Between." In *Marco Polo and the Encounter of East and West*, edited by Suzanne Akbari and Amilcare Iannucci, 3–20. Toronto: University of Toronto Press, 2008.

AKIPress. "Kazakh-Chinese Business Council Sign Deals Worth $10bn." *Eurasian Business Briefing*, December 15, 2015. http://www.eurasianbusinessbriefing.com/10bn-worth-of-deals-signed-at-kazakh-chinese-business-council/.

Al Azm, Amr. "The Pillaging of Syria's Cultural Heritage." Middle East Institute. May 22, 2015. http://www.mei.edu/content/at/pillaging-syrias-cultural-heritage.

Ali, Mohamed Azmin. "The Maritime Silk Road and a New Paradigm for Globalisation." Speech at 5th Malaysia-China Entrepreneur Conference, Malaysia, November 7, 2015. http://azminali.com/speech-the-5th-malaysia-china-entreprenur-conference/.

Allison, Graham. *Destined for War: Can America and China Escape Thucydides's Trap?* London: Scribe, 2017. Kindle.

Almond, Roncevert Ganan. "Summits, Roads and Suspended Disbelief in Central Asia." *The Diplomat*, June 27, 2017. http://thediplomat.com/2017/06/summits-roads-and-suspended-disbelief-in-central-asia/.

Alram, Michael. "The History of the Silk Road as Reflected in Coins." *Parthica* 6 (2004): 47–68.

Anderson, Benedict. *Imagined Communities: Reflections on the Origin and Spread of Nationalism*. Ithaca, NY: Cornell University Press, 1991.

Aneja, Atul. "China Developing Soft-Power Infra along Silk Road." *The Hindu*, October 18, 2016. http://www.thehindu.com/news/international/China-developing-soft-power-infra-along-Silk-Road/article14308346.ece.

Appadurai, Arjun. "Introduction: Commodities and the Politics of Value." In *The Social Life of Things: Commodities in Cultural Perspective*, edited by Arjun Appadurai, 3–63. Cambridge: Cambridge University Press, 1988.

Arase, David. "China's Two Silk Roads Initiative: What It Means for Southeast Asia." *Southeast Asian Affairs*, no.1 (2015): 25–45.

Asia Pacific Exchange and Cooperation Foundation. "Silk Road Dunhuang International Music Forum Starts." July 26, 2015. http://www.apecf.org/en/foundationnews/20150726.html.

Australian National University. "China Maritime Silk Road Museum." *China Heritage Newsletter* (China Heritage Project, Australian National University). March 28, 2016. http://www.chinaheritagequarterly.org/articles.php?searchterm=001_maritimesilk.inc&issue=001.

Aydin, Cemil. *The Politics of Anti-Westernism in Asia: Visions of World Order in Pan-Islamic and Pan-Asian Thought*. New York: Columbia University Press, 2007. Kindle.

Babbage, Ross. *Countering China's Adventurism in the South China Sea: Strategy Options for the Trump Administration*. Washington, DC: Center for Strategic and Budgetary Assessments, 2016. http://csbaonline.org/research/publications/countering-chinas-adventurism-in-the-south-china-sea-strategy-options-for-t.

Baedeker, Karl. *Russia, with Teheran, Port Arthur, and Peking: Handbook for Travellers*. Leipzig: Karl Baedeker, 1914.

Barfield, Thomas. "Steppe Empires, China, and the Silk Route: Nomads as a Force in International Trade and Politics." In *Nomads in the Sedentary World*, edited by Anatoly Khazanov and Andre Wink, 234–49. Surrey, UK: Curzon Press, 2001.

Barnard, Anne. "Saudi's Grant to Lebanon Is Seen as Message to U.S." *New York Times*, January 6, 2014. http://www.nytimes.com/2014/01/07/world/middleeast/saudis-grant-to-lebanon-is-seen-as-message-to-us.html?_r=0.

Barnes, Ruth. "Yami Boats and Boat Building in a Wider Perspective." In *Ships and the Development of Maritime Technology on the Indian Ocean*, edited by David Parkin and Ruth Barnes, 291–314. London: Routledge, 2016. Kindle.

Barry, Andrew. *Material Politics: Disputes along the Pipeline*. Oxford, UK: Wiley Blackwell, 2013.

———. *Political Machines: Governing a Technological Society*. London: A & C Black, 2001.

Bayly, Christopher. *The Birth of the Modern World, 1780–1914: Global Connections and Comparisons*. Malden, MA: Blackwell, 2012.

BBC News. "China Invests $124bn in Belt and Road Global Trade Project." May 14, 2017. http://www.bbc.com/news/world-asia-39912671.

Beardsley, Brent, Jorge Becerra, Federico Burgoni, Bruce Holley, Daniel Kessler, Federico Muxi, Matthias Naumann, Tjun Tang, André Xavier, and Anna Zakrzewski. "Global Wealth 2015: Winning the Growth Game." June 15, 2015. https://www.bcgperspectives.com/content/articles/financial-institutions-growth-global-wealth-2015-winning-the-growth-game/?chapter=2#chapter2_section2.

Beaujard, Philippe. "The Worlds of the Indian Ocean." In *Trade, Circulation, and Flow in the Indian Ocean World*, edited by Michael Pearson, 15–26. Houndmills, UK: Palgrave Macmillan, 2015. Kindle.

———. *L'Océan Indien, au cœur des globalisations de l'Ancien Monde (7e-15e siècles)*. Paris: Collins, 2012.

Beaumont, Joan. "The Diplomacy of Extra-territorial Heritage: The Kokoda Track, Papua New Guinea." *International Journal of Heritage Studies* 22, no. 5 (2016): 355–67.

Becker, Seymour. "The 'Great Game': The History of an Evocative Phrase." *Asian Affairs* 43, no. 1 (2012): 61–80.

Beckerleg, Susan. "Continuity and Adaptation by Contemporary Swahili Boatbuilders in Kenya." In *Ships and the Development of Maritime Technology on the Indian Ocean*, edited by David Parkin and Ruth Barnes, 259–77. London: Routledge, 2016. Kindle.

Beckert, Sven. *Empire of Cotton: A New History of Global Capitalism*. London: Penguin, 2015.

Beckwith, Christopher. *Empires of the Silk Road: A History of Central Eurasia from the Bronze Age to the Present*. Princeton, NJ: Princeton University Press, 2009. Kindle.

Bell, Daniel. "Realising Tianxia: Traditional Values and China's Foreign Policy." In *Chinese Visions of World Order: Tianxia, Culture, and World Politics*, edited by Ban Wang, 129–46. Durham, NC: Duke University Press, 2017. Kindle.

Best Restaurants of Australia. "Silk Road." April 20, 2017. https://www.bestrestaurants.com.au/vic/melbourne/melbourne-cbd/restaurant/silk-road.

Bevan, Robert. *The Destruction of Memory: Architecture at War*. London: Reaktion Books, 2006.

Bhoothalingam, Ravi. "The Silk Road as a Global Brand." *China Report* 52, no. 1 (2016): 45–52.

Bhutta, Zafar. "Chinese Offer to Finance Whole $2bn LNG Project." *Express Tribune*, May 14, 2016. http://tribune.com.pk/story/1103009/lng-terminal-and-pipeline-chinese-offer-to-finance-whole-2-billion-project/.

Biran, Michal. "Introduction: Nomadic Culture." In *Nomads as Agents of Cultural Change: The Mongols and Their Eurasian Predecessors*, edited by Leuven Amitai and Michal Biran, 1–9. Honolulu: University of Hawai'i Press, 2015.

Blackburn, Anne. *Locations of Buddhism: Colonialism and Modernity in Sri Lanka*. Chicago: University of Chicago Press, 2010.

Blank, Stephen. "The Dynamics of Russo-Chinese Relations." In *The Routledge Handbook of Asian Security Studies*, edited by Sumit Ganguly, Andrew Scobell, and Joseph Liow Chin Long, 74–88. London: Routledge, 2018. Kindle.

Bohrer, Frederick. *Photography and Archaeology*. London: Reaktion Books, 2011.

Bonavia, Judy. *Collins Illustrated Guide to the Silk Road*. London: Collins, 1988.

Bouchard, Christian, and William Crumplin. "Neglected No Longer: The Indian Ocean at the Forefront of World Geopolitics and Global Geostrategy." *Journal of the Indian Ocean* 6, no. 1 (2010): 26–51.

Boulnois, Luce. *The Silk Road*. London: George Allen and Unwin, 1966.

Bourdieu, Pierre. *The Field of Cultural Production: Essays on Art and Literature*. New York: Columbia University Press, 1993.

Braudel, Fernand. *Civilization and Capitalism: 15th–18th century*. Vol. 1. Berkeley: University of California Press, 1992.

———. *The Mediterranean and the Mediterranean World in the Age of Philip II*. Vol. 1. Berkeley: University of California Press, 2012.

Brewster, David. "Beyond the 'String of Pearls': Is There Really a Sino-Indian Security Dilemma in the Indian Ocean?" *Journal of the Indian Ocean Region* 10, no. 2 (2014): 133–49.

———. "An Indian Ocean Dilemma: Sino-Indian Rivalry and China's Strategic Vulnerability in the Indian Ocean." *Journal of the Indian Ocean Region* 11, no. 1 (2015): 48–59.

British Library. "The International Dunhuang Project." http://idp.bl.uk.

Bromby, Robin. "Down the Maritime Silk Road." *The Australian*, December 6, 2013. http://www.theaustralian.com.au/business/in-depth/down-the-maritime-silk-road/story-fnjy4qn5-1226776242929.

Brook, Timothy, Michael Van Walt van Praag, and Miek Boltjes, eds. *Sacred Mandates: Asian International Relations since Chinggis Khan*. Chicago: University of Chicago Press, 2018. Kindle.

Brown, Rachel. "Where Will the New Silk Road Lead? The Effects of Chinese Investment and Migration in Xinjiang and Central Asia." *Columbia University Journal of Politics and Society* 26 (2016): 69–91.

Bueger, Christian. "Actor-Network Theory, Methodology, and International Organization." *International Political Sociology* 7, no. 3 (2013): 338–42.

Bulkin, V., Leo Klejn, and G. S. Lebedev. "Attainments and Problems of Soviet Archaeology." *World Archaeology* 13, no. 3 (1982): 272–95.

Byrne, Denis. "The Problem with Looting: An Alternative Perspective on Antiquities Trafficking in Southeast Asia." *Journal of Field Archaeology* 41, no. 3 (2016): 344–54.

Calder, Kent. *The New Continentalism: Energy and Twenty-First-Century Eurasian Geopolitics*. New Haven, CT: Yale University Press, 2012.

Callahan, William. "China's 'Asia Dream': The Belt Road Initiative and the New Regional Order." *Asian Journal of Comparative Politics* 1, no. 3 (2016): 226–43.

———. *China: The Pessoptimist Nation*. Oxford: Oxford University Press, 2010. Kindle.

Cambodia Daily. "Cambodia Positions Itself along New Silk Road." June 27, 2016. https://www.cambodiadaily.com/news/cambodia-positions-itself-along-new-silk-road-114629/.

Carissimo, Justin. "Isis Propaganda Video Shows 25 Syrian Soldiers Executed by Teenage Militants in Palmyra." *The Independent*, July 5, 2015. https://www.independent.co.uk/news/world/middle-east/isis-propaganda-video-shows-20-syrian-government-soldiers-executed-in-palmyra-by-young-islamists-10366533.html.

Castells, Manuel, and Gustavo Cardoso, eds. *The Network Society: From Knowledge to Policy*. Washington, DC: Johns Hopkins Center for Transatlantic Relations, 2005.

Channel NewsAsia. "China, Iran Vow Tighter Ties as Xi Visits." January 23, 2016. http://www.channelnewsasia.com/news/asia/china-iran-vow-tighter-ties-as-xi-visits-8212976.

Chaudhuri, Kirti. *Asia before Europe: Economy and Civilisation of the Indian Ocean from the Rise of Islam to 1750*. Cambridge: Cambridge University Press, 1990.

Chew, Amy. "China, Malaysia Tout New 'Port Alliance' to Reduce Customs Bottlenecks and Boost Trade." *South China Morning Post*, April 9, 2016. http://www.scmp.com/news/asia/southeast-asia/article/1934839/china-malaysia-tout-new-port-alliance-reduce-customs.

Chew, Sing. "The Southeast Asian Connection in the First Eurasian World Economy, 200BCE–CE 500." In *Trade, Circulation, and Flow in the Indian Ocean*, edited by Michael Pearson, 27–54. Houndmills, UK: Palgrave Macmillan, 2015. Kindle.

Chin, Tamara. "The Invention of the Silk Road, 1877." *Critical Inquiry* 40, no. 1 (2013): 194–219.

ChinaDaily. "Quanzhou to Build Maritime Silk Road Museum." January 5, 2016. http://www.chinadaily.com.cn/m/fujian/jinjiang/2016–01/05/content_22959273.htm.

"China Lends $300m to Build Strategic Road in Kyrgyzstan." Alexander's Gas & Oil Connections: An Institute for Global Energy Research. http://www.gasandoil.com/oilaround/2015/06/china-lends-300-mm-to-build-strategic-road-in-kyrgyzstan.

China National Silk Museum. "'Silk on the Silk Road' Exhibition Unveils in Cairo." http://en.chinasilkmuseum.com/news/detail_604.html.

———. "'Silk on the Silk Road' Exhibition Unveils in Luxor." http://en.chinasilkmuseum.com/news/detail_605.html.

———. "'Silk on the Silk Road'—Chinese Silk Exhibition Unveils in Qatar." http://en.chinasilkmuseum.com/news/detail_603.html.

China.org.cn. "Xi Tours 'Living Fossil of Silk Road' Bukhara in Uzbekistan." June 22, 2016. http://www.china.org.cn/world/2016–06/22/content_38716368.htm.

China Underwater Cultural Heritage. *National Conservation Center for Underwater Cultural Heritage*. http://www.unesco.org/new/fileadmin/MULTIMEDIA/HQ/CLT/pdf/CUCH_brochure_en (2)_02.pdf.

Chinese Academy of Social Sciences. "Maritime Silk Road on Show in Fujian." Chinese Archaeology: Institute of Archaeology Chinese Academy of Social Sciences. http://www.kaogu.net.cn/en/News/Academic_activities/2014/1212/48579.html.

Chowdhary, Mahwish. "China's Billion-Dollar Gateway to the Subcontinent: Pakistan May Be Opening a Door It Cannot Close." *Forbes*, August 25, 2015. http://www.forbes.com/sites/realspin/2015/08/25/china-looks-to-pakistan-to-expand-its-influence-in-asia/#11189f987179.

Chulov, Martin. "How an Arrest in Iraq Revealed Isis's $2bn Jihadist Network." *The Guardian*, June 16, 2014. https://www.theguardian.com/world/2014/jun/15/iraq-isis-arrest-jihadists-wealth-power.

Church, Sally, John Gebhardt, and Terry Little. "A Naval Architectural Analysis of the Plausibility of 450-ft Treasure Ships." In *Zheng He and the Afro-Asian World*, edited by Chia Lin Sien and Sally Kathryn Church, 15–47. Melaka: Perbadanan Muzium Melaka and International Zheng He Society, 2012.

City University of Hong Kong. "Introduction." City University of Hong Kong—Research Center on One-Belt-One-Road. May 2017. http://www.cb.cityu.edu.hk/obor/.

City University of Macau. "Macau 'One Belt, One Road' Research Center." City University of Macau. May 2017. http://www.cityu.edu.mo/moborrc/en/home.

Clarke, Amy. "Heritage Diplomacy." In *Handbook of Cultural Security*, edited by Yasushi Watanabe, 417–36. Cheltenham, UK: Edward Elgar Publishing, 2018.

Clarke, Michael. "The Belt and Road Initiative: China's New Grand Strategy?" *Asia Policy* 24 (2017): 71–79.

———. "China and the Uyghur: The 'Palestinization' of Xinjiang." *Middle East Policy* 22, no. 3 (2015): 127–46.

———. *Xinjiang and China's Rise in Central Asia—A History*. London: Routledge, 2011.

———. "Xinjiang from the 'Outside-In' and the 'Inside-Out': Exploring the Imagined Geopolitics of a Contested Region." In *Inside Xinjiang: Space,*

Place and Power in China's Muslim Far Northwest, edited by Anna Hayes and Michael Clarke, 225–59. Oxford: Routledge, 2016.

Cline, Eric. *Biblical Archaeology: A Very Short Introduction*. Oxford: Oxford University Press, 2009. Kindle.

Coastweek.com. "Kenya Islands Show Footprint of Chinese Expeditionary Voyages." http://www.coastweek.com/3828-Kenyan-islands-with-footprint-of-Chinese-expeditionary-voyages.htm.

Collins, Randall. "Civilizations as Zones of Prestige and Social Contact." *International Sociology* 16, no. 3 (2001): 421–37.

Collins, Robert. *East to Cathay: The Silk Road*. New York: McGraw-Hill Book Co., 1986.

Cooper, Andrew, Jorge Heine, and Ramesh Thakur, eds. *The Oxford Handbook of Modern Diplomacy*. Oxford: Oxford University Press, 2013.

Cox, Simon. "The Men Who Smuggle the Loot That Funds IS." *BBC News*, February 17, 2015. http://www.bbc.com/news/magazine-31485439.

Dabashi, Hamid. *Iran without Borders: Toward a Critique of the Postcolonial Nation*. London: Verso, 2016.

Daily Star. "China to Build 2 More Bridges in Bangladesh." August 26, 2015. http://www.thedailystar.net/country/china-build-2-more-bridges-bangladesh-132925.

Davis, Thomas. *Shifting Sands: The Rise and Fall of Biblical Archaeology*. Oxford: Oxford University Press, 2004. Kindle.

Dawn. "Iran, China Agree $600-Billion Trade Deal after Sanctions." January 23, 2016. http://www.dawn.com/news/1234923.

Deleuze, Giles, and Félix Guattari. *A Thousand Plateaus: Capitalism and Schizophrenia*. London: Bloomsbury Academic, 2013.

Demattè, Paola. "Emperors and Scholars: Collecting Culture and Late Imperial Antiquarianism." In *Collecting China: The World, China, and a Short History of Collecting*, edited by Vimalin Rujivacharakul, 165–75. Newark: University of Delaware Press, 2011.

Desmarais, France, ed. *Countering Illicit Traffic in Cultural Goods: The Global Challenge of Protecting the World's Heritage*. Paris: International Council of Museums International Observatory on Illicit Traffic in Cultural Goods, 2015.

Devendra, Somasiri. "Pre-Modern Sri Lankan Ships." In *Ships and the Development of Maritime Technology on the Indian Ocean*, edited by David Parkin and Ruth Barnes, 128–73. London: Routledge, 2016. Kindle.

Devonshire-Ellis, Chris. *China's New Economic Silk Road: The Great Eurasian Game and the String of Pearls*. Hong Kong: Asia Briefing, Dezan Shira and Associates, 2015.

Ding Yi. "Huaqiao University—A Think Tank for 'One Belt, One Road' Strategy." *Sino-US.com*, June 10, 2015. http://www.sino-us.com/433/15021790078.html.

Dolukhanov, Pavel. "Archaeology and Nationalism in Totalitarian and Post-Totalitarian Russia." In *Nationalism and Archaeology: Scottish Archaeological*

Forum, edited by John Atkinson, Iain Banks, and Jerry O'Sullivan, 200–213. Glasgow: Cruithne Press, 1996.

Duara, Prasenjit. "The Chinese World Order and Planetary Sustainability." In *Chinese Visions of World Order: Tianxia, Culture and World Politics*, edited by Ban Wang, 65–83. Durham, NC: Duke University Press, 2017. Kindle.

Dündar, Ali Merthan. "The Effects of the Russo-Japanese War on the Turkic Nations: Japan and Japanese in Folk Songs, Elegies and Poems." In *Japan on the Silk Road: Encounters and Perspectives of Politics and Culture in Eurasia*, edited by Selçuk Esenbel, 199–227. Leiden: Brill, 2018.

The Economist. "Mad about Museums." January 6, 2014. http://www.economist.com/news/special-report/21591710-china-building-thousands-new-museums-how-will-it-fill-them-mad-about-museums.

———. "A Russian Orchestra Plays Bach and Prokofiev in the Ruins of Palmyra." May 6, 2016. https://www.economist.com/europe/2016/05/06/a-russian-orchestra-plays-bach-and-prokofiev-in-the-ruins-of-palmyra.

Edwards, Penny. *Cambodge: The Cultivation of a Nation, 1860–1945*. Honolulu: University of Hawai'i Press, 2007.

Egami, Namio. "The Silk Road and Japan." In *The Sea Route: The Grand Exhibition of Silk Road Civilizations*, edited by Silk Road Exposition, 10–20. Nara: Nara National Museum, 1988.

Elisseef, Vadime. *The Silk Roads: Highways of Culture and Commerce*. New York: Berghahn Books, 2000.

Embassy of the People's Republic of China in the Islamic Republic of Pakistan. "China-Pakistan Friendship: As Pure and Sincere as the Ever-Flowing Water." Embassy of the People's Republic of China in the Islamic Republic of Pakistan. November 25, 2015. http://pk.china-embassy.org/eng/zbgx/t1318449.htm.

Erickson, Andrew, and Kevin Bond. "Archaeology and the South China Sea." *The Diplomat*, July 20, 2015. http://thediplomat.com/2015/07/archaeology-and-the-south-china-sea/.

Errington, Elizabeth, and Vesta Curtis. "The British and Archaeology in Nineteenth-Century Persia." In *From Persepolis to the Punjab: Exploring Ancient Iran, Afghanistan and Pakistan*, edited by Elizabeth Errington and Vesta Curtis, 166–78. London: British Museum Press, 2007.

———. "The Explorers and Collectors." In *From Persepolis to the Punjab: Exploring Ancient Iran, Afghanistan and Pakistan*, edited by Elizabeth Errington and Vesta Curtis, 3–16. London: British Museum Press, 2007.

Errington, Shelly. *The Death of Authentic Primitive Art and Other Tales of Progress*. Berkeley: University of California Press, 1998.

Esenbel, Selçuk. Introduction to *Japan on the Silk Road: Encounters and Perspectives of Politics and Culture in Eurasia*, edited by Selçuk Esenbel, 1–34. Leiden: Brill, 2018.

Fagan, Brian. *Return to Babylon: Travelers, Archaeologists, and Monuments in Mesopotamia*. Boulder: University of Colorado Press, 2007.

Fatemi, Syed Tariq. "Speech by Special Assistant to the Prime Minister Syed Tariq Fatemi at the Karachi Council of Foreign Relations." Speech at Karachi Council on Foreign Relations, Islamabad, February 2, 2015. http://www.mofa.gov.pk/chile/pr-details.php?prID=2546.

Ferdinand, Peter. "Westward Ho? The China Dream and 'One Belt, One Road': Chinese Foreign Policy under Xi Jinping." *International Affairs* 92, no. 4 (2016): 941–57.

Fernando, M. "Continuity and Change in Maritime Trade in the Straits of Melaka in the Seventeenth and Eighteenth Centuries." In *Trade, Circulation, and Flow in the Indian Ocean World,* edited by Michael Pearson, 109–28. Houndmills, UK: Palgrave Macmillan, 2015. Kindle.

Financial Express. "BD, China Sign Deal to Build SEZ in Ctg." May 3, 2017. http://www.thefinancialexpress-bd.com/2016/06/17/34490/BD,-China-sign-deal-to-build-SEZ-in-Ctg.

Finlay, Robert. *The Pilgrim Art: Cultures of Porcelain in World History.* Berkeley: University of California Press, 2010. Kindle.

Flecker, Michael. *The Archaeological Excavation of the 10th Century Intan Shipwreck.* Oxford: Archaeopress, 2002.

———. "A Ninth-Century Arab or Indian Shipwreck in Indonesia: The First Archaeological Evidence of Direct Trade with China." In *Shipwrecked: Tang Treasures and Monsoon Winds,* edited by Regina Krahl and Alison Effeny, 101–19. Singapore: Arthur M. Sackler Gallery, Smithsonian Institution, and National Heritage Board, Singapore Tourism Board, 2010.

———. "The Thirteenth-Century Java Sea Wreck: A Chinese Cargo in an Indonesian Ship." *Mariner's Mirror* 89, no. 4 (2003): 388–404.

Flood, Finbarr, and Gülru Necipoğlu. *A Companion to Islamic Art and Architecture.* New York: John Wiley and Sons, 2017. Kindle.

Foltz, Richard. *Religions of the Silk Road: Premodern Patterns of Globalization.* New York: Springer Publishing, 2010.

Forsby, Andrea. "An End to Harmony? The Rise of a Sino-centric China." *Political Perspectives* 5, no. 3 (2011): 5–26.

Frank, Andre Gunder. *ReOrient: Global Economy in the Asian Age.* Berkeley: University of California Press, 1998.

Frankopan, Peter. *The Silk Roads: A New History of the World.* London: Bloomsbury, 2015.

French, Howard. *Everything Under the Heavens: How the Past Helps Shape China's Push for Global Power.* New York: Knopf, 2017. Kindle.

Frost, Mark Ravinder. "Handing Back History: Britain's Imperial Heritage State in Colonial Sri Lanka and South Asia, 1870–1920." Keynote address at National Symposium of Historical Studies, University of Sri Lanka, Sri Lanka, January 31, 2018.

Galambos, Imre. "Buddhist Relics from the Western Regions: Japanese Archaeological Exploration of Central Asia." In *Writing Travel in Central Asian*

History, edited by Nile Green, 152–69. Bloomington: Indiana University Press, 2014. Kindle.

———. "Japanese Exploration of Central Asia: The Ōtani Expeditions and Their British Connections." *Bulletin of SOAS* 75, no. 1 (2012): 113–34.

———. "Japanese 'Spies' along the Silk Road: British Suspicions Regarding the Second Ōtani (1908–09)." *Japanese Religions* 35, no. 1–2 (2010): 33–61.

Gallagher, Steven. "'Purchased in Hong Kong': Is Hong Kong the Best Place to Buy Stolen or Looted Antiquities?" *International Journal of Cultural Property* 24 (2017): 479–96.

Gannon, Megan. "600 Year Old Coin Found in China." *LiveScience*, March 14, 2013. http://www.livescience.com/27890-chinese-coin-found-in-kenya.html.

Gayamov, A. "Soviet Music." *Soviet Travel* 3 (1933).

GiziMap. *Silk Road Countries*. Budapest: GiziMap, 2015.

Gladney, Dru. "The Party-State's Nationalist Strategy to Control the Uyghur: Silenced Voices." In *Routledge Handbook of the Chinese Communist Party*, edited by Willy Wo-Lap Lam, 311–31. Oxford: Routledge, 2018. Kindle.

Godahewa, Nalaka. "Commemorating Zheng He, the Greatest Navigator to Visit Sri Lanka from China." *Daily FT*, February 4, 2012. http://www.ft.lk/article/69932/Commemorating-Zheng-He—the-greatest-navigator-to-visit-Sri-Lanka-from-China.

Golombek, Lisa. "Dominant Fashions and Distinctive Styles." In *Persian Pottery in the First Global Age: The Sixteenth and Seventeenth Centuries*, edited by Lisa Golombek, Robert Mason, Patricia Proctor, and Eileen Reilly, 57–122. Leiden: Brill, 2014.

Golombek, Lisa, and Eileen Reilly. "Safavid Society and the Ceramic Industry." In *Persian Pottery in the First Global Age: The Sixteenth and Seventeenth Centuries*, edited by Lisa Golombek, Robert Mason, Patricia Proctor, and Eileen Reilly, 13–56. Leiden: Brill, 2014.

Government of Georgia. "PM Discusses Georgia-China Partnership at Beijing University." September 11, 2015. http://gov.ge/index.php?lang_id=ENG&sec_id=412&info_id=51429.

Government of the Hong Kong Special Administrative Region. "Museum of History Exhibition Reveals Development of China's Maritime Silk Road (with photos)." Press release. October 25, 2016. http://www.info.gov.hk/gia/general/201610/25/P2016102500265.htm.

GovTrack. "H.R. 2867—105th Congress: Silk Road Strategy Act of 1997." June 26, 2018. https://www.govtrack.us/congress/bills/105/hr2867.

Green, Nile. "Introduction: Writing, Travel, and the Global History of Central Asia." In *Writing Travel in Central Asian History*, edited by Nile Green, 1–40. Bloomington: Indiana University Press, 2014. Kindle.

Griffiths, John. *Tea: A History of the Drink That Changed the World*. London: Andre Deutsch, 2011.

Guha-Thakurta, Tapati. *Monuments, Objects, Histories: Institutions of Art in Colonial and Postcolonial India*. New York: Columbia University Press, 2004.

Habib, Talha Bin. "China to Help Build 2nd Runway, 3rd Terminal at HSIA." *Financial Express*, July 4, 2013. http://print.thefinancialexpress-bd.com/old/index.php?ref=MjBfMDdfMDRfMTNfMV85MF8xNzUyNzU=.

Hahn, Hans Peter, and Hadas Weiss, eds. *Mobility, Meaning and the Transformation of Things (Shifting Contexts of Material Culture through Time and Space)*. Oxford, UK: Oxbow Books, 2013. Kindle.

Hamashita, Takeshi. *China, East Asia and the Global Economy: Regional and Historical Perspectives*. London: Routledge, 2008.

Hanban News. "The First 'Confucius Institute of Maritime Silk Road' Established in Thailand." July 6, 2015. http://english.hanban.org/article/2015-07/06/content_608534.htm.

Hannerz, Ulf. "Geocultural Scenarios." In *Frontiers of Sociology*, edited by Peter Hedstrom and Bjorn Wittrock, 267–88. Leiden: Brill, 2009.

Hansen, Valerie. *The Silk Road: A New History with Documents*. Oxford: Oxford University Press, 2017.

Harilela, Divia. "Hong Kong Fashion Designer Vivienne Tam Talks about Her First Public Show of Art." *South China Morning Post*, April 15, 2016. http://www.scmp.com/lifestyle/fashion-luxury/article/1936128/hong-kong-fashion-designer-vivienne-tam-talks-about-her.

Harrell, Stevan, ed. *Cultural Encounters on China's Ethnic Frontiers*. Seattle: University of Washington, 1995. Kindle.

Harrison, Rodney. *Understanding the Politics of Heritage*. Manchester: Manchester University Press, 2010.

He Yini. "China to Invest $900b in Belt and Road Initiative." *China Daily USA*, May 28, 2015. http://usa.chinadaily.com.cn/business/2015-05/28/content_20845687.htm.

Hedin, Sven. *The Silk Road*. London: George Routledge and Sons, 1938.

Heine, Jorge. "From Club to Network Diplomacy." In *The Oxford Handbook of Modern Diplomacy*, edited by Andrew Cooper, Jorge Heine, and Ramesh Thakur, 54–69. Oxford: Oxford University Press, 2013.

Herlijanto, Johanes. "Cultivating the Past, Imagining the Future: Enthusiasm for Zheng He in Contemporary Indonesia." In *Zheng He and the Afro-Asian World*, edited by Chia Lin Sien and Sally Kathryn Church, 130–46. Melaka: Perbadanan Muzium Melaka and International Zheng He Society, 2012.

———. "What Does Indonesia's Pribumi Elite Think of Ethnic Chinese Today?" *ISEAS Perspective* 2016, no. 32 (2016): 1–9.

The Hindu. "Celebrating 60 Years of Diplomatic Ties." March 24, 2016. http://www.thehindu.com/news/cities/Kochi/celebrating-60-years-of-diplomatic-ties/article7947457.ece.

Hirsch, Francine. *Empire of Nations: Ethnographic Knowledge and the Making of the Soviet Union*. Ithaca, NY: Cornell University Press, 2014. Kindle.

Hobson, John. *The Eastern Origins of Western Civilisation*. Cambridge: Cambridge University Press, 2004.

Hodder, Ian. *Entangled: An Archaeology of the Relationships between Humans and Things*. Malden, MA: Wiley-Blackwell, 2012.

Hofmeyer, Isabel. "Styling Multilateralism: Indian Ocean Cultural Futures." *Journal of the Indian Ocean Region* 11, no. 1 (2015): 98–109.

Holmes, James, and Toshi Yoshihara. "Soft Power Goes to Sea." *American Interest* (Spring 2008): 66–67.

Hopkirk, Peter. *Foreign Devils on the Silk Road: The Search for the Lost Treasures of Central Asia*. London: John Murray, 1989. Kindle.

Honey, William Bower. *The Ceramic Art of China and Other Countries of the Far East*. London: Faber and Faber, 1945.

Horowitz, Shale, and Peng Yu. "Holding China's West: Explaining CCP Strategies of Rule in Tibet and Xinjiang." *Journal of Chinese Political Science* 20, no. 4 (2015): 451–75.

Houghton, Arthur. *Dark Athena*. N.p.: CreateSpace Independent Publishing Platform, 2016.

Howarth, Dan. "China 'Can't Buy Culture' With Museum Boom, Say Critics." *Dezeen*. December 11, 2015. http://www.dezeen.com/2015/12/11/new-chinese-museums-construction-boom-opening-money-cant-buy-culture-china/.

Hu Jintao. "Enhance China-Africa Unity and Cooperation to Build a Harmonious World." Speech at University of Pretoria, Pretoria, February 7, 2007. http://www.fmprc.gov.cn/mfa_eng/wjdt_665385/zyjh_665391/t298174.shtml.

———. "Speech by Chinese President Hu Jintao given to Federal Parliament." Speech at Australian Parliament House, Canberra, October 24, 2003. http://www.smh.com.au/articles/2003/10/24/1066631618612.html.

Hvistendahl, Mara. "Rebuilding a Treasure Ship." *Archaeology* 61, no. 2 (2008): 40–45.

Iannucci, Amilcare, and John Tulk. "From Alterity to Holism: Cinematic Depictions of Marco Polo and His Travels." In *Marco Polo and the Encounter of East and West*, edited by Suzanne Conklin Akbari and Amilcare Iannucci, 201–43. Toronto: University of Toronto Press, 2008.

I Cross China. "'Maritime Silk Road' Themed Subway Hits Ningbo." May 21, 2015. http://www.icrosschina.com/sizzling/2015/0521/12875.shtml.

Ingram, Edward. "Great Britain's Great Game: An Introduction." *International History Review* 2, no. 2 (1980): 160–71.

Inozemtsev, Vladislav. "Why Kazakhstan Holds the Keys to the Global Economy." *The Independent*, November 10, 2015. http://www.independent.co.uk/voices/why-kazakhstan-holds-the-keys-to-the-global-economy-a6727391.html.

Institute of Developing Economies—Japan External Trade Organization. "China in Africa." http://www.ide.go.jp/English/Data/Africa_file/Manualreport/cia_10.html.

International Council of Museums. "Fighting Illicit Traffic." http://icom.museum/programmes/fighting-illicit-traffic/.

International Crisis Group. *Stirring Up the South China Sea (II): Regional Responses*. Brussels: International Crisis Group, 2012.

Irvine, Roger. "Nationalists or Internationalists? China's International Relations Experts Debate the Future." *Journal of Contemporary China* 26, no. 106 (2017): 586–600.

Jacobs, Justin. "Confronting Indiana Jones: Chinese Nationalism, Historical Imperialism and the Criminalisation of Aurel Stein and the Raiders of Dunhuang, 1899–1944." In *China on the Margins*, edited by Sherman Cochran and Paul Pickowicz, 65–90. Ithaca, NY: Cornell University East Asia Program, 2010.

———. "Cultural Thieves or Political Liabilities? How Chinese Officials Viewed Foreign Archaeologists in Xinjiang, 1839–1914." *Silk Road* 10 (2012): 117–22.

———. "Nationalist China's 'Great Game': Leveraging Explorers in Xinjiang, 1927–1935." *Journal of Asian Studies* 73, no. 1 (2014): 43–64.

———. *Xinjiang and the Modern Chinese State*. Studies on Ethnic Groups in China. Seattle: University of Washington Press, 2016. Kindle.

Japan Consortium for International Cooperation in Cultural Heritage. "The Project on the Archaeological Research Project on the Sites of Palmyra." http://www.jcic-heritage.jp/en/project/middle_east_palmyra_201109/.

Japanese National Commission for UNESCO, ed. *Research in Japan in History of Eastern and Western Cultural Contacts: Its Development and Present Situation*. Tokyo: Japanese National Commission for UNESCO, 1957.

Jiang Zemin. "Enhance Mutual Understanding and Build Stronger Ties of Friendship and Cooperation." Speech at Harvard University, Cambridge, November 1, 1997. http://www.china-embassy.org/eng/zmgx/zysj/jzxfm/t36252.htm.

Joniak-Luthi, Agneishka. "Roads in China's Borderlands: Interfaces of Spatial Representations, Perceptions, Practices, and Knowledges." *Modern Asian Studies* 50, no. 1 (2016): 118–40.

Kane, Eileen. *Russian Hajj: Empire and the Pilgrimage to Mecca*. Ithaca, NY: Cornell University Press, 2015. Kindle.

Kariyawasam, Prasad. "Sri Lanka—A Hub in the Indian Ocean." Speech at the East West Center, Hawai'i, February 11, 2016. http://slembassyusa.org/embassy_press_releases/remarks-by-ambassador-prasad-kariyawasam-at-the-east-west-center-hawaii-sri-lanka-a-hub-in-the-indian-ocean/.

Kennedy, Richard. "The Silk Road: Connecting Cultures, Creating Trust." Smithsonian Folklife Festival. https://festival.si.edu/2002/the-silk-road/the-silk-road-connecting-cultures-creating-trust/smithsonian.

Khanna, Parag. *Connectography: Mapping the Global Network Revolution*. London: Weidenfeld and Nicolson, 2016.

Khazanov, Anatoly. "Nomads in the History of the Sedentary World." In *Nomads in the Sedentary World*, edited by Anatoly Khazanov and Andre Wink, 1–23. Surrey, UK: Curzon Press, 2001.

Khorana, Sangeeta, and Leila Choukroune. "India and the Indian Ocean Region." *Journal of the Indian Ocean Region* 12, no. 2 (2016): 122–25.

Kiernan, Matthew. "Silk Road Key to China's Next Move." *Huffington Post*, August 31, 2015. http://www.huffingtonpost.com/matthew-j-kiernan/silk-road-key-to-chinas-n_b_8068318.html.

Kingsley, Sean. "Editorial: Tang Treasures, Monsoon Winds and a Storm in a Teacup." *The Undertow* (blog). March 13, 2011. https://wreckwatch.org/2011/03/13/editorial-tang-treasures-monsoon-winds-and-a-storm-in-a-teacup/.

Kivimäki, Timo. "Soft Power and Global Governance with Chinese Characteristics." *Chinese Journal of International Politics* 7, no. 4 (2014): 421–77.

Klejn, Leo. *Soviet Archaeology: Schools, Trends and History*. Oxford: Oxford University Press, 2012.

Kong, Lily, and Justin O'Connor. *Creative Economies, Creative Cities: Asian-European Perspectives*. New York: Springer, 2009.

Kopytoff, Igor. "The Cultural Biography of Things: Commoditization." In *The Social Life of Things: Commodities in Cultural Perspective*, edited by Arjun Appadurai, 64–94. Cambridge: Cambridge University Press, 1988.

Korybko, Andrew. "What Could India's 'Cotton Route' Look Like?" *Sputnik International*, March 29, 2015. http://sputniknew.com/columnists/20150329/1020170021.html.

Krahl, Regina, and Alison Effeny, eds. *Shipwrecked: Tang Treasures and Monsoon Winds*. Singapore: Arthur M. Sackler Gallery, Smithsonian Institution, and National Heritage Board, Singapore Tourism Board, 2010.

Krishnan, Ananth. "China Offers to Develop Chittagong Port." *The Hindu*, March 15, 2010. http://www.thehindu.com/news/international/china-offers-to-develop-chittagong-port/article245961.ece.

Küçükyalçin, Erdal. "Ōtani Kozui and His Vision of Asia: From Villa Nirakusō to 'The Rise of Asia' project." In *Japan on the Silk Road: Encounters and Perspectives of Politics and Culture in Eurasia*, edited by Selçuk Esenbel, 181–98. Leiden: Brill, 2018.

Kurlansky, Mark. *Paper: Paging through History*. New York: W. W. Norton and Co., 2017.

Kuzmina, Elena. *The Prehistory of the Silk Road*. Edited by Victor H. Mair. Philadelphia: University of Pennsylvania, 2008.

Kwa Chong-Guan, ed. *Early Southeast Asia Viewed from India: An Anthology of Articles from the "Journal of the Greater India Society."* Delhi: Manohar, 2013.

Langfitt, Frank. "China Builds Museums, but Filling Them Is Another Story." *All Things Considered*, May 21, 2013. http://www.npr.org/sections/parallels/2013/05/21/185776432/china-builds-museums-but-will-the-visitors-come.

Latour, Bruno. *Reassembling the Social: An Introduction to Actor-Network-Theory*. Oxford: Oxford University Press, 2007.

Law, John. *After Method: Mess in Social Science Research*. London: Routledge, 2004.

Lawton, John, ed. *The Integral Study of the Silk Roads: Roads of Dialogue*. Paris: UNESCO, 2008.

Leibold, James. "Interethnic Conflict in the PRC: Xinjiang and Tibet as Exceptions?" In *Protest and Xinjiang: Unrest in China's West*, edited by Ben Hillman and Gary Tuttle, 223–50. New York: Columbia University Press, 2016.

———. *Reconfiguring Chinese Nationalism: How the Qing Frontier and Its Indigenes Became Chinese*. New York: Palgrave Macmillan, 2007. Kindle.

Li Rongxia. "'Significance of Zheng He's Voyages.'" *Beijing Review*, n.d. http://www.bjreview.cn/EN/En-2005/05–28-e/china-2.htm.

Lim Tai Wei. Introduction to *China's One Belt One Road Initiative*, edited by Lim Tai Wei, Henry Chan Hing Lee, Katherine Tseng Hui-Yi, and Lim Wen Xi, 3–18. London: Imperial College Press, 2016.

Lim Tai Wei, Henry Chan Hing Lee, Katherine Tseng Hui-Yi, and Lim Wen Xi, eds. *China's One Belt One Road Initiative*. London: Imperial College Press, 2016.

Liu, Xinru. *The Silk Road in World History*. Oxford: Oxford University Press, 2010. Kindle.

Lord, Gail, and Ngaire Blankenberg, eds. *Cities, Museums and Soft Power*. Washington, DC: American Alliance of Museums, 2015.

Lord, Gail, and Ngaire Blankenberg. "Introduction: Why Cities, Museums and Soft Power." In *Cities, Museums and Soft Power*, edited by Gail Lord and Ngaire Blankenberg, 20–24. Washington, DC: American Alliance of Museums, 2015.

Lu, Tracey L. D. *Museums in China: Materialized Power and Objectified Identities*. London: Routledge, 2014.

Luke, Christina, and Morag Kersel. *U.S. Cultural Diplomacy and Archaeology: Soft Power, Hard Heritage*. New York: Routledge, 2013.

Mackinder, Halford. "The Geographical Pivot of History." *Geographical Journal* 23, no. 4 (1904): 421–37.

Mahfud, Choirul. "Lessons from Zheng He: Love of Peace and Multiculturalism." In *Zheng He and the Afro-Asian World*, edited by Chia Lin Sien and Sally Kathryn Church, 188–91. Melaka: Perbadanan Muzium Melaka and International Zheng He Society, 2012.

———. "The Role of Cheng Ho Mosque: The New Silk Road, Indonesia-China Relations in Islamic Cultural Identity." *Journal of Indonesian Islam* 8, no. 1 (2014): 31–34.

Majlesi, Afshin. "Tehran Hosts Sino-Iranian Friendship Conference as Ancient Silk Road Backs to Life." *Tehran Times*, October 24, 2016. http://www.tehrantimes.com/news/407630/Tehran-hosts-Sino-Iranian-friendship-conference-as-ancient-Silk.

Mallawarachi, Bharatha. "Sri Lanka Signs on to China's Maritime 'Silk Road.'" *Sydney Morning Herald*, September 17, 2014. http: http://www.smh.com.au/

world/sri-lanka-signs-on-to-chinas-maritime-silk-road-20140916–10hx4a.html.

Manzara Tourism. "WTO and the Silk Road." http://www.manzaratourism.com/gsr_wto.

Marchand, Suzanne. *German Orientalism in the Age of Empire: Religion, Race, and Scholarship*. Cambridge: Cambridge University Press, 2009.

Maritime Silk Road Society. "Leaders from Top-notch Enterprises Speak at 'The Belt and Road Initiative—Combining Hard Power with Soft Power' Conference." February 27, 2016. http://maritimesilkroad.org.hk/en/news/newsReleasesDetail/102.

Martin, Terry. *The Affirmative Action Empire: Nations and Nationalism in the Soviet Union, 1923–1939*. Ithaca, NY: Cornell University Press, 2017. Kindle.

Matsuda, Hisao. "General Survey: The Development of Researches in the History of the Intercourse between East and West in Japan." In *Research in Japan in History of Eastern and Western Cultural Contacts: Its Development and Present Situation*, edited by Japanese National Commission for UNESCO, 1–18. Tokyo: Japanese National Commission for UNESCO, 1957.

Mauss, Marcel. *The Gift*. London: Routledge, 2002.

Mavin, Duncan. "Calculating the Revenue from Antiquities to Islamic State." *Wall Street Journal*, February 11, 2015. http://www.wsj.com/articles/calculating-the-revenue-from-antiquities-to-islamic-state-1423657578.

Mawdsley, Emma. "Development Geography 1: Cooperation, Competition and Convergence between 'North' and 'South.'" *Progress in Human Geography* 41, no. 1 (2017): 108–17.

Mayor, Federico. "Preface of the Director General of UNESCO." In *The Integral Study of Silk Roads: Roads of Dialogue*, edited by John Lawton, 3. Paris: UNESCO, 2008.

Meere, Isabelle. "Asian Leaders Clash over Belt and Road at SCO Summit." Center of Expertise on Asia. June 12, 2018. https://asiahouse.org/asian-leaders-clash-belt-road-sco-summit/.

Meskell, Lynn. "The Rush to Inscribe: Reflections on the 35th Session of the World Heritage Committee UNESCO Paris, 2011." *Journal of Field Archaeology* 37, no. 2 (2012): 145–51.

———. "States of Conservation: Protection, Politics, and Pacting within UNESCO's World Heritage Committee." *Anthropological Quarterly* 87, no. 1 (2014): 217–43.

Meskell, Lynn, Claudia Liuzza, Enrico Bertacchini, and Donatella Saccone. "Multilateralism and UNESCO World Heritage: Decision-Making, States Parties and Political Processes." *International Journal of Heritage Studies* 21, no. 5 (2015): 423–40.

Meyer, Karl, and Sharon Brysac. *Tournament of Shadows: The Great Game and the Race for Empire in Central Asia*. New York: Basic Books, 1999.

Mignolo, Walter. *The Darker Side of the Renaissance: Literacy, Territoriality, and Colonization*. Ann Arbor: University of Michigan Press, 2003.

Miksic, John. Introduction to *Ancient Silk Trade Routes: Selected Works from Symposium on Cross Cultural Exchanges and Their Legacies in Asia*, edited by Qin Dashu and Jian Yuan, 1–17. Singapore: World Scientific, 2014.

———. *Singapore and the Silk Road of the Sea, 1300–1800*. Singapore: National University of Singapore Press, 2013.

Millward, James. *Eurasian Crossroads: A History of Xinjiang*. London: Hurst and Co., 2007.

———. "Uyghur Art Music and the Ambiguities of Chinese Silk Roadism in Xinjiang." *Silkroad Foundation Newsletter*, June 2005. http://www.silkroadfoundation.org/newsletter/v013num1/3_uyghur.php.

———. *The Silk Road: A Very Short Introduction*. Oxford: Oxford University Press, 2013. Kindle.

———. "What It's Like to Live in a Surveillance State." *New York Times*, February 3, 2018. https://www.nytimes.com/2018/02/03/opinion/sunday/china-surveillance-state-uighurs.html.

Milner, Anthony. "Culture and the International Relations of Asia." *Pacific Review* 30, no. 6 (2017): 857–69.

Ministry of Foreign Affairs, Islamic Republic of Afghanistan. *The Sixth Regional Economic Cooperation Conference on Afghanistan Report*. N.p.: Ministry of Foreign Affairs Islamic Republic of Afghanistan, 2015.

Ministry of Foreign Affairs of the People's Republic of China. "Joint Statement on Strengthening Comprehensive Strategic Partnership between the People's Republic of China and the Republic of Indonesia." March 27, 2015. https://www.fmprc.gov.cn/mfa_eng/wjdt_665385/2649_665393/t1249201.shtml.

———. "President Xi Jinping Delivers Important Speech at Pakistan's Parliament Entitled 'Building a China-Pakistan Community of Shared Destiny to Pursue Closer Win-Win Cooperation.'" April 21, 2015. http://www.fmprc.gov.cn/mfa_eng/topics_665678/xjpdbjstjxgsfwbfydnxycxyfldrhyhwlhy60znjnhd/t1257288.shtml.

———. "Xi Jinping Delivers Important Speech at Headquarters of the League of Arab States, Stressing to Jointly Create a Bright Future for Development of China-Arab Relations and Promote National Rejuvenation of China and Arab States to Form More Convergence." January 22, 2016. http://www.fmprc.gov.cn/mfa_eng/topics_665678/xjpdstajyljxgsfw/t1334587.shtml.

———. "Xi Jinping Meets with President Maithripala Sirisena of Sri Lanka." Ministry of Foreign Affairs of the People's Republic of China. October 17, 2016. http://www.fmprc.gov.cn/mfa_eng/zxxx_662805/t1406780.shtml.

Ministry of Foreign Affairs, Sri Lanka. "Admiral Zheng He Promotes Tourism to Sri Lanka from the Port City of Taicang." January 14, 2013. http://www.mfa.gov.lk/index.php/missions/mission-activities/3817-admiral-zheng-he-promotes-tourism-to-sri-lanka-from-the-port-city-of-taicang.

———. "Joint Statement between the People's Republic of China and the Democratic Socialist Republic of Sri Lanka at the Conclusion of the Official Visit of Prime Minister Ranil Wickremesinghe." April 9, 2016. http://

www.mfa.gov.lk/index.php/en/media/media-releases/6452-jointstatement-slpmvisitchina.

Miskimmon, Alister, Ben O'Loughlin, and Laura Roselle, eds. *Strategic Narratives: Communication Power and the New World Order.* London: Routledge, 2013. Kindle.

Moallem, Minoo. *Persian Carpets: The Nation as a Transnational Commodity.* Oxford: Routledge, 2018. Kindle.

Mohd, Aminuk Karim. "The South China Sea Disputes: Is High Politics Overtaking?" *Pacific Focus* 28, no. 1 (2013): 99–119.

Moody, Oliver. "Isis Fills War Chest by Selling Looted Antiquities to the West." *The Times*, December 17, 2014. http://www.thetimes.co.uk/tto/news/world/middleeast/article4299572.ece.

Morgan, Joyce, and Conrad Walters. *Journeys on the Silk Road.* Guilford, NY: Lyons Press, 2012.

Mori, Masao. "The Steppe Route." In *Research in Japan in History of Eastern and Western Cultural Contacts: Its Development and Present Situation*, edited by Japanese National Commissions for UNESCO, 19–34. Tokyo: Japanese National Commission for UNESCO, 1957.

Morris, Rosalind. *Photographies East: The Camera and Its Histories in East and Southeast Asia.* Durham, NC: Duke University Press, 2009.

Mostert, Tristan, and Jan van Campen. *Silk Thread: China and the Netherlands from 1600.* Nijmegen, The Netherlands: Uitgeverij Vantilt, 2015.

Mozaffari, Ali. *Forming National Identity in Iran: The Idea of Homeland Derived from Ancient Persia and the Islamic Imaginations of Peace.* London: I. B. Tauris, 2014. Kindle.

Murphy, Zoe. "Zheng He: Symbol of China's 'Peaceful Rise.'" *BBC News*, July 28, 2010. http://www.bbc.com/news/world-asia-pacific-10767321.

Murray, Stuart, Paul Sharp, Geoffrey Wiseman, David Criekemans, and Jan Melissen. "The Present and Future of Diplomacy and Diplomatic Studies." *International Studies Review* 13, no. 4 (2011): 709–28.

Myrdal, Jan. *The Silk Road: A Journey from the High Pamirs and Ili through Sinkiang and Kansu.* London: Victor Gollancz, 1980.

Nader, Sami. "Saudi Arabia, France Make Their Move in Lebanon." *Al-Monitor*, January 6, 2014. http://www.al-monitor.com/pulse/originals/2014/01/saudi-france-partnership-lebanon-iran.html.

Nathan, Gary. *Cumin, Camels and Caravans: A Spice Odyssey.* Berkeley: University of California Press, 2014.

Nathan, Sellapan Ramanathan. "Speech by HE President SR Nathan, President of Singapore, Welcoming Jewel of Muscat to Singapore." Speech at ceremony to celebrate the arrival of the Jewel of Muscat to Singapore, Singapore, July 3, 2010. http://jewelofmuscat.tv/speeches/speech-by-he-president-sr-nathan-president-of-singapore/.

National Development and Reform Commission—People's Republic of China. "Vision and Actions on Jointly Building Silk Road Economic Belt and

21st-Century Maritime Silk Road, Issued by the National Development and Reform Commission, Ministry of Foreign Affairs, and Ministry of Commerce of the People's Republic of China, with State Council authorization." March 28, 2015. http://en.ndrc.gov.cn/newsrelease/201503/t20150330_669367.html.

National Marine Sanctuary Foundation and University of Hawai'i. "Final APCONF 2014 Program." Asia-Pacific Conference on Underwater Cultural Heritage. http://www.apconf.org/wp-content/uploads/Final-APCONF-2014-program.pdf.

News.gov.hk. "Treasure Trade: Maritime Silk Road." November 20, 2016. http://www.news.gov.hk/en/city_life/html/2016/11/20161117_154216.shtml.

Nexon, Daniel, and Vincent Pouliot. "Things of Networks: Situating ANT in International Relations." *International Political Sociology* 7, no. 3 (2013): 342–45.

Nguyen, Hong Thao. "Vietnam's Position on the Sovereignty over the Paracels and the Spratlys: Its Maritime Claims." *Journal of East Asia and International Law* 5, no. 1 (2012): 165–211.

Nish, Ian. "Japan and the Great Game." In *Japan on the Silk Road: Encounters and Perspectives of Politics and Culture in Eurasia*, edited by Selçuk Esenbel, 35–47. Leiden: Brill, 2018.

Nye, Joseph. *The Future of Power*. New York: Public Affairs, 2011.

O'Hara, Sarah. "Great Game or Grubby Game? The Struggle for Control of the Caspian." *Geopolitics* 9, no. 1 (2004): 138–60.

Olenin, Boris. "Sukhum-Kaleh, City of Joy." *Soviet Travel* 3 (1933): 15–17.

Ollier, Leakthina Chau-Pech, and Tim Winter, eds. *Expressions of Cambodia: The Politics of Tradition, Identity and Change*. London: Routledge, 2006.

Ondaatjie, Anusha. "China Maritime Silk Road Is Sri Lanka's Boon as Xi Visits." *Bloomberg*, September 17, 2014. https://www.bloomberg.com/news/articles/2014-09-15/china-maritime-silk-road-proves-boon-for-sri-lanka-as-xi-arrives.

Pan, Su-Yan, and Joe Tin-yau Lo. "Re-conceptualizing China's Rise as a Global Power: A Neo-Tributary Perspective." *Pacific Review* 30, no. 1 (2017): 1–25.

Parkinson, Joe, Ayla Albayrak, and Duncan Mavin. "Syrian 'Monuments Men' Race to Protect Antiquities as Looting Bankrolls Terror." *Wall Street Journal*, February 10, 2015. http://www.wsj.com/articles/syrian-monuments-men-race-to-protect-antiquities-as-looting-bankrolls-terror-1423615241.

Parzinger, Hermann. "The 'Silk Roads' Concept Reconsidered: About Transfers, Transportation and Transcontinental Interactions in Prehistory." *Silk Road* 5, no. 2 (2008): 7–15.

Pattanaik, Smruti. "Indian Ocean in the Emerging Geo-Strategic Context: Examining India's Relations with Its Maritime South Asian Neighbours." *Journal of the Indian Ocean Region* 12, no. 2 (2016): 126–42.

Paviour, Ben. "For China, 'Cambodia Is a Sideshow, but It's a Loyal One.'" *Cambodia Daily*, October 19, 2016. https://www.cambodiadaily.com/news/china-cambodia-sideshow-loyal-one-119475/.

Pearson, Michael. *The Indian Ocean*. London: Routledge, 2003.

———. "Introduction: Maritime History and the Indian Ocean World." In *Trade, Circulation, and Flow in the Indian Ocean World*, edited by Michael Pearson, 1–14. Houndmills, UK: Palgrave Macmillan, 2015. Kindle.

Peleggi, Maurizio. *Thailand: The Worldly Kingdom*. London: Reaktion Books, 2007.

Peyrouse, Sebastien. "The Evolution of Russia's Views on the Belt and Road Initiative." *Asia Policy* 24 (2017): 96–102.

Pfister, Rudolph. *Nouveaux textiles de Palmyre*. Paris: Les éditions d'art et d'historie, 1937.

———. *Textiles de Palmyre*. Paris: Les éditions d'art et d'historie, 1934.

———. *Textiles de Palmyre III*. Paris: Les éditions d'art et d'historie, 1940.

Pierson, Stacey. *From Object to Concept: Global Consumption and the Transformation of Ming Porcelain*. Hong Kong: Hong Kong University Press, 2014.

Plets, Gertjan. "Ethno-nationalism, Asymmetric Federalism and Soviet Perceptions of the Past: (World) Heritage Activism in the Russian Federation." *Journal of Social Archaeology* 15, no. 1 (2015): 67–93.

———. "Heritage Bureaucracies and the Modern Nation State: Towards an Ethnography of Archaeological Systems of Government." *Archaeological Dialogues* 23, no. 2 (2016): 193–213.

———. "Violins and Trowels for Palmyra: Post-Conflict Heritage Politics." *Anthropology Today* 33, no. 4 (2017): 18–22.

Poling, Gregory. *Grappling with the South China Sea Policy Challenge*. Washington, DC: Center for Strategic and International Studies, 2015.

Pomeranz, Kenneth. *The Great Divergence: China, Europe and the Making of the Modern World Economy*. Princeton, NJ: Princeton University Press, 2001.

Porter, Benjamin. "Near Eastern Archaeology: Imperial Pasts, Postcolonial Presents, and the Possibilities of a Decolonized Future." In *Handbook of Postcolonial Archaeology*, edited by Jane Lydon and Uzma Z. Rizvi, 51–60. Oxford: Routledge, 2016.

Press TV. "Chabahar Port to Harbor Chinese Industrial Town." April 27, 2016. http://www.presstv.com/Detail/2016/04/27/462797/China-CMI-Iran-mega-port.

Proctor, Patricia. "The Measure of Faithfulness: The Chinese Models for Safavid Blue and White." In *Persian Pottery in the First Global Age: The Sixteenth and Seventeenth Centuries*, edited by Lisa Golombek, Robert Mason, Patricia Proctor, and Eileen Reilly, 123–68. Leiden: Brill, 2014.

PSCU. "Kenya and China Sign Sh42bn Lamu Port Deal." *Daily Nation*, August 1, 2014. http://www.nation.co.ke/counties/nairobi/Kenya-and-China-Lamu-Port-deal/1954174-2405458-9yviq5/index.html.

Qu Zhe. "Speech on the Belt and Road Initiative by Chine Ambassador Qu Zhe." Speech at Chinese Embassy in Estonia, Estonia, April 13, 2013. http://www.fmprc.gov.cn/mfa_eng/wjb_663304/zwjg_665342/zwbd_665378/t1254259.shtml.

Ramachandran, Sudha. "Iran, China and the Silk Road Train." *The Diplomat*, March 20, 2016. http://thediplomat.com/2016/03/iran-china-and-the-silk-road-train/.

Ray, Himanshu. "Seafaring in Peninsular India in the Ancient Period of Watercraft and Maritime Communities." In *Ships and the Development of Maritime Technology on the Indian Ocean*, edited by David Parkin and Ruth Barnes, 62–91. London: Routledge, 2016. Kindle.

Reeves, Jeffrey. "Origins, Intentions, and Security Implications of Xi Jinping's Belt and Road Initiative." In *The Routledge Handbook of Asian Security Studies*, edited by Sumit Ganguly, Andrew Scobell, and Joseph Liow Chin Long, 61–73. London: Routledge, 2018. Kindle.

Reid, Anthony. *Southeast Asia in the Age of Commerce 1450–1680*. Vol. 2, *Expansion and Crisis*. New Haven, CT: Yale University Press, 1993.

Rezakhani, Khodadad. "The Road That Never Was: The Silk Road and Trans-Eurasian Exchange." *Comparative Studies of South Asia, Africa and the Middle East* 30, no. 3 (2010): 420–33.

Rieber, Alfred. *The Struggle for the Eurasian Borderlands: From the Rise of Early Modern Empires to the End of the First World War.* Cambridge: Cambridge University Press, 2014. Kindle.

Rogers, Roy, and Jatswan Sidhu. "International Norms and Human Rights Conditions in the Xinjiang Uyghur Autonomous Region (XUAR)." *Malaysian Journal of International Relations* 4 (2017): 109–37.

Rolland, Nadège. *China's Eurasian Century? Political and Strategic Implications of the Belt and Road Initiative*. Seattle: National Bureau of Asian Research, 2017.

Russell-Smith, Lilla. "Hungarian Explorers in Dunhuang." *Journal of the Asiatic Society* 10, no. 3 (2000): 341–62.

Ruta, Michele, and Mauro Boffa. "Trade Linkages among Belt and Road Economies: Three Facts and One Prediction." *The Trade Post*. May 31, 2018. https://blogs.worldbank.org/trade/trade-linkages-among-belt-and-road-economies-three-facts-and-one-prediction.

Ryono, Angel, and Matthew Galway. "Xinjiang under China: Reflections on the Multiple Dimensions of the 2009 Urumqi Uprising." *Asian Ethnicity* 16, no. 2 (2015): 235–55.

Saales, Sven, and Christopher Szpilman. *Pan-Asianism: A Documentary History, Vol. 1: 1850–1920*. Lanham, MD: Rowman and Littlefield, 2011. Kindle.

Sarkisova, Oksana. *Screening Soviet Nationalities: Kulturfilms from the Far North to Central Asia*. London: I. B. Tauris, 2017. Kindle.

Sassen, Saskia. *Global Networks, Linked Cities*. London: Routledge, 2002.

Schaffer, Teresita, and Howard Schaffer. *India at the Global High Table: The Quest for Regional Primacy and Strategic Autonomy*. Uttar Pradesh, India: Harper Collins, 2016.

Schatzki, Theodore. "Materiality and Social Life." *Nature and Culture* 5, no. 2 (2010): 123–49.

Schöttli, Jivanta. "Special Issue: Power, Politics and Maritime Governance in the Indian Ocean." *Journal of the Indian Ocean Region* 9, no. 1 (2013): 1–5.

Scobell, Andrew. "Whither China's 21st Century Trajectory?" In *The Routledge Handbook of Asian Security Studies*, edited by Sumit Ganguly, Andrew Scobell, and Joseph Liow Chin Long, 11–20. London: Routledge, 2018. Kindle.

Selbitschka, Armin. "The Early Silk Road(s)." In *Oxford Encyclopedia of Asian History*, edited by David Ludden. New York: Oxford University Press, 2018. http://asianhistory.oxfordre.com/view/10.1093.acrefore/9780190277727.001.0001/acrefore-9780190277727–2–2.

Sen, Tansen. *Buddhism, Diplomacy and Trade: The Realignment of India-China Relations, 600–1400*. Lanham, MD: Rowman & Littlefield, 2016. Kindle.

———. "Diplomacy, Trade and the Quest for the Buddha's Tooth: The Yongle Emperor and Ming China's South Asian Frontier." In *Ming China: Courts and Contacts, 1400–1450C*, edited by Jessica Harrison-Hall, Craig Clunas, and Luk Yu-ping, 26–36. London: British Museum, 2016.

———. "The Impact of Zheng He's Expeditions on Indian Ocean Interactions." *Bulletin of the School of Oriental and African Studies* 79, no. 3 (2016): 609–36.

———. "Silk Road Diplomacy—Twists, Turns and Distorted History." YaleGlobal Online. September 23, 2014. http://yaleglobal.yale.edu/content/silk-road-diplomacy-twists-turns-and-distorted-history.

Shah, Saeed. "China to Build Pipeline from Iran to Pakistan." *Wall Street Journal*, April 9, 2015. http://www.wsj.com/articles/china-to-build-pipeline-from-iran-to-pakistan-1428515277.

Shambaugh, David. *China Goes Global: The Partial Power*. Oxford: Oxford University Press, 2013.

Share, Michael. "The Great Game Revisited: Three Empires Collide in Chinese Turkestan." *Europe-Asia Studies* 67, no. 7 (2015): 1102–29.

Shepard, Wade. "China's Jewel in the Heart of the Indian Ocean." *The Diplomat*, May 9, 2016. http://thediplomat.com/2016/05/chinas-jewel-in-the-heart-of-the-indian-ocean/.

Sheriff, Abdul. "The Dhow Culture of the Western Indian Ocean." In *Journeys and Dwellings: Indian Ocean Themes in South Asia*, edited by Helene Basu, 61–89. Hyderabad, India: Orient Blackswan, 2016. Kindle.

Silberman, Neil. "Promised Lands and Chosen Peoples: The Politics and Poetics of Archaeological Narrative." In *Nationalism, Politics and the Practice of Archaeology*, edited by Philip Kohl and Clare Fawcett, 120–38. Cambridge: Cambridge University Press, 1996.

Silk Road Exposition, ed. *The Oasis and Steppe Routes: The Grand Exhibition of Silk Road Civilizations*. Nara: Nara National Museum, 1988.

———. *The Route of Buddhist Art: The Grand Exhibition of Silk Road Civilizations*. Nara: Nara National Museum, 1988.

———. *The Sea Route: The Grand Exhibition of Silk Road Civilizations*. Nara: Nara National Museum, 1988.

Silkroad Project. "Silkroad Musicians." http://www.silkroadproject.org/ensemble.

Silk Road Universities Network. "Constitution." August 22, 2015. http://www.sun-silkroadia.org/_en/sun/constitutions.php.

———. "Purpose and Background." http://www.sun-silkroadia.org/_en/sun/purpose.php.

Sims, Kearrin. "The Asian Development Bank and the Production of Poverty: Neoliberalism, Technocratic Modernization and Land Dispossession in the Greater Mekong Subregion." *Singapore Journal of Tropical Geography* 36, no. 1 (2015): 112–26.

Singh, Antara. "China's Soft Power Projection across the Oceans." *Maritime Affairs: Journal of the National Maritime Foundation of India* 12, no. 1 (2016): 25–37.

———. "India, China and the US: Strategic Convergence in the Indo-Pacific." *Journal of the Indian Ocean Region* 12, no. 2 (2016): 161–76.

Smithsonian Folklife Festival. "The Silk Road Ensemble." https://festival.si.edu/2002/the-silk-road/the-silk-road-ensemble/smithsonian.

Snodgrass, Judith. *Presenting Japanese Buddhism to the West: Orientalism, Occidentalism, and the Columbian exposition*. Chapel Hill: University of North Carolina Press, 2003.

South China Morning Post. "After 'Chinese Dream,' Xi Jinping Outlines Vision for 'Asia-Pacific Dream' at APEC Meet." November 10, 2014. http://www.scmp.com/news/china/article/1635715/after-chinese-dream-xi-jinping-offers-china-driven-asia-pacific-dream.

Stargardt, Janice. "Indian Ocean Trade in the Ninth and Tenth Centuries: Demand, Distance, and Profit." *South Asian Studies* 30, no. 1 (2014): 35–55.

Starr, S. Frederick. Introduction to *The New Silk Roads: Transport and Trade in Greater Central Asia*, edited by S. Frederick Starr, 5–32. Washington, DC: Central Asia-Caucasus Institute, 2007.

———. *Lost Enlightenment: Central Asia's Golden Age from the Arab Conquest to Tamerlane*. Princeton, NJ: Princeton University Press, 2013. Kindle.

Starr, S. Frederick, ed. *Xinjiang: China's Muslim Borderland*. Armonk, NY: M. E. Sharpe, 2004.

St. Clair, Michael. *The Great Chinese Art Transfer: How So Much of China's Art Came to America*. Madison, NJ: Fairleigh Dickinson University Press, 2016.

Stevens, Stuart. *Night Train to Turkistan: Adventures along China's Silk Road*. London: Paladin Grafton Books, 1990.

Stone, Peter, Joanne Farchakh Bajjaly, and Robert Fisk. *The Destruction of Cultural Heritage in Iraq*. Woodbridge, UK: Boydell Press, 2008.

Strauch, Ingo. "Priority and Exclusiveness: Russians and Germans at the Northern Silk Road (Materials from the Turfan-Akten)." *Études de lettres* 2–3 (2014): 147–78.

Sudakova, Elena, ed. *See USSR: Intourist Posters and the Marketing of the Soviet Union*. London: GRAD Publishing, 2013.

Swenson, Astrid, and Peter Mandler, eds. *From Plunder to Preservation: Britain and the Heritage of the Empire, c. 1800–1940*. Oxford: Oxford University Press, 2013.

Sylvester, Christine. *Art/Museums: International Relations Where We Least Expect It*. Boulder, CO: Paradigm Publishers, 2009.

Tan, See Seng, and Amitav Acharya, eds. *Bandung Revisited: The Legacy of the 1955 Asian-African Conference for International Order*. Singapore: NUS Press, 2008.

Tang West Market Group. "TWMG's Contribution to the Revival of the Silk Road of UNESCO." http://www.tangwestmarket.com/2015/09/twmgs-contribution-to-the-revival-of-the-silk-road-of-unesco-3.

Tan Ta Sen. "Cheng Ho's Guanchang Site in Melaka." In *Zheng He and the Afro-Asian World*, edited by Chia Lin Sien and Sally Kathryn Church, 192–215. Melaka: Perbadanan Muzium Melaka and International Zheng He Society, 2012.

———. "Cheng Ho Spirit and World Dream." In *China's One Belt One Road Initiative*, edited by Lim Tai Wei, Henry Chan Hing Lee, Katherine Tseng Hui-Yi, and Lim Wen Xin, 57–59. London: Imperial College Press, 2016.

Tankha, Brij. "Exploring Asia, Reforming Japan: Ōtani and Itō Chūta." In *Japan on the Silk Road: Encounters and Perspectives of Politics and Culture in Eurasia*, edited by Selçuk Esenbel, 156–80. Leiden: Brill, 2018.

Thang, Nguyen-Dang, and Hong Thao Nguyen. "China's Nine Dotted Lines in the South China Sea: The 2011 Exchange of Diplomatic Notes between the Philippines and China." *Ocean Development and International Law* 43, no. 1 (2012): 35–56.

Thaw, Lawrence, and Margaret Thaw. "Along the Old Silk Routes." *National Geographic Magazine*, 78, 1940.

Thorsten, Marie. "Silk Road Nostalgia and Imagined Global Community." *Comparative American Studies: An International Journal* 3, no. 3 (2005): 301–17.

Thum, Rian. "How Should the World Respond to Intensifying Repression in Xinjiang." *ChinaFile*, June 4, 2018. http://www.chinafile.com/conversation/how-should-world-respond-intensifying-repression-xinjiang.

Tiezzi, Shannon. "Chinese Company Wins Contract for Deep Sea Port in Myanmar." *The Diplomat*, January 1, 2016. http://thediplomat.com/2016/01/chinese-company-wins-contract-for-deep-sea-port-in-myanmar/.

Ting, Joseph. *The Maritime Silk Route: 2000 Years of the South China Sea*. Hong Kong: Urban Council, 1996.

Trigger, Bruce. *A History of Archaeological Thought*. Cambridge: Cambridge University Press, 1989.

Trigkas, Vasilis. "The New Silk Road: Cultural and Economic Diplomacy China-Greece-Europe." Tsinghua University. September 26, 2014. http://www.imir.tsinghua.edu.cn/publish/iisen/7259/2014/20141010134153540200352/20141010134153540200352_.html.

Tseng Hui-Yi, Katherine. "The South China Sea and the Maritime Silk Road Proposal: Conflicts Can Be Transformed." In *China's One Belt One Road Initiative*, edited by Lim Tai Wei, Henry Chan Hing Lee, Katherine Tseng Hui-Yi, and Lim Wen Xin, 133–47. London: Imperial College Press, 2016.

Tsing, Anna. *Friction: An Ethnography of Global Connection.* Princeton, NJ: Princeton University Press, 2005. Kindle.
UNESCO. *Adoption of Retrospective Statements of Outstanding Universal Value.* WHC-10/34.COM/8E. Paris: UNESCO, 2010. https://whc.unesco.org./archive/2010/whc10-34Com-9Be.pdf.
———. *Convention Concerning the Protection of the World Cultural and Natural Heritage.* Paris: UNESCO, 1972. http://whc.unesco.org/en/conventiontext.
———. "Historic Centre of Bukhara." UNESCO World Heritage Centre. 2016. http://whc.unesco.org/en/list/602.
———. *Integral Study of the Silk Roads: Roads of Dialogue, 1998–1997.* Paris: UNESCO, 2008. http://unesdoc.unesco.org/images/0015/001591/159189E.pdf.
———. "Mausoleum of Khoja Ahmed Yasawi." UNESCO World Heritage Centre. http://whc.unesco.org/en/list/1103.
———. "The Operational Guidelines for the Implementation of the World Heritage Convention." UNESCO World Heritage Centre. 2016. http://whc.unesco.org/pg.cfm?cid=57.
———. *Report on Serial Nominations and Properties.* Paris: UNESCO, 2010. http://whc.unesco.org/archive/2010/whc10–34Com-9Be.pdf.
———. "Silk Roads World Heritage Serial and Transnational Nomination in Central Asia: A UNESCO/Japanese Funds-in-Trust Project." UNESCO World Heritage Centre. http://whc.unesco.org/en/activities/825/.
———. "Support for the Silk Roads World Heritage Sites in Central Asia (Phase II)." UNESCO World Heritage Centre. http://whc.unesco.org/en/activities/870.
UNESCO Kazakhstan. "At UNESCO, Kazakhstan's President Nazarbayev Calls for Intercultural Dialogue to Counter Extremism." http://www.unesco.kz/new/en/unesco/news/2998.
———. "UNESCO and 'The Centre for the Rapprochement of Cultures' in Kazakhstan Have Defined an Action Plan for the Intercultural Dialogue." August 5, 2016. http://en.unesco.kz/the-first-scientific-and-methodological-meeting-of-the-council-of-the-state-museum-of-the.
UNESCO and Republic of Korea Funds-in-Trust-Project. *Support for the Preparation for the World Heritage Serial Nomination of the Silk Roads in South Asia.* Paris: UNESCO, 2016. http://unesdoc.unesco.org/images/0024/002460/246096e.pdf.
UNESCO TV and NHK Nippon Hoso Kyokai. "UNESCO/NHK Videos on Heritage." https://whc.unesco.org/en/list/23/video.
UN World Tourism Organization. *Silk Road Action Plan 2016/7.* Madrid: UNWTO, 2017. http://silkroad.unwto.org/node/30284.
Varutti, Marzia. *Museums in China: The Politics of Representation after Mao.* Woodbridge, UK: Boydell and Brewer, 2014.
Vernet, Thomas. "East African Travelers and Traders in the Indian Ocean: Swahili Ships, Swahili Mobilities, ca. 1500–1800." In *Trade, Circulation,*

and Flow in the Indian Ocean World, edited by Michael Pearson, 167–202. Houndmills, UK: Palgrave Macmillan, 2015. Kindle.

Vickers, Edward. "A Civilizing Mission with Chinese Characteristics? Education, Colonialism and Chinese State Formation In Comparative Perspective." In *Constructing Modern Asian Citizenship*, edited by Edward Vickers and Krishna Kumar, 50–79. Oxford: Routledge, 2015.

Vlasic, Mark, and Helga Turku. "'Blood Antiquities.'" *Journal of International Criminal Justice* 14, no. 5 (2016): 1043–1300.

Volker, Tijs. *Porcelain and the Dutch East India Company: As Recorded in the Dagh-Registers of Batavia Castle, Those of Hirado and Deshima and Other Contemporary Papers, 1602–1682*. Leiden: Brill, 1954.

Voll, John Obert. "Main Street of Eurasia." Silk Roads: Dialogue, Diversity & Development. https://en.unesco.org/silkroad/content/main-street-eurasia.

Volmer, John, E. J. Keall, and E. Nagai-Berthrong. *Silk Roads—China Ships: An Exhibition of East-West Trade*. Toronto: Royal Ontario Museum, 1983.

Wade, Geoff. "Chinese Engagement with the Indian Ocean during the Song, Yuan and Ming Dynasties (Tenth to Sixteenth Centuries)." In *Trade, Circulation, and Flow in the Indian Ocean World*, edited by Michael Pearson, 55–81. Houndmills, UK: Palgrave Macmillan, 2015. Kindle.

———. "Civilisational Rhetoric and the Obfuscation of Power Politics." In *Sacred Mandates: Asian International Relations since Chinggis Khan*, edited by Timothy Brook, Michael van Walt van Praag, and Miek Boltjes, 75–89. Chicago: University of Chicago Press, 2018. Kindle.

———. "The 'Native Office' System: A Chinese Mechanism for Southern Territorial Expansion over Two Millennnia." In *Asian Expansions: The Historical Experiences of Polity Expansion in Asia*, edited by Geoff Wade, 69–90. London: Routledge, 2015.

———. "The Zheng He Voyages: A Reassessment." ARI Working Paper No. 31, Asia Research Institute, Singapore, October 2004.

Wake, Christopher. "The Myth of Zheng He's Great Treasure Ships." *International Journal of Maritime History* 16, no. 1 (2004): 59–76.

Walker, Michael. "Romano-Indian Rouletted Pottery in Indonesia." *Asian Perspectives: The Bulletin of the Far-Eastern Prehistory Association* 20, no. 2 (1980): 228–35.

Wallerstein, Immanuel. *Geopolitics and Geoculture: Essays on the Changing World-System*. Cambridge: Cambridge University Press, 1991.

———. *The Modern World-System I: Capitalist Agriculture and the Origins of the European World-Economy in the Sixteenth Century*. Berkeley: University of California Press, 2011.

———. *The Modern World System II: Mercantilism and the Consolidation of the European World-Economy 1600–1750*. Berkeley: University of California Press, 2011.

———. *The Modern World-System III: The Second Era of Great Expansion of the Capitalist World-Economy, 1730s–1840s*. Berkeley: University of California Press, 2011.

Wang, Ban. "The Moral Vision in Kong Youwei's Book of the Great Community." In *Chinese Visions of World Order: Tianxia, Culture, and World Politics*, edited by Ban Wang, 87–105. Durham, NC: Duke University Press, 2017. Kindle.

Wang, Helen. "Sir Aurel Stein." In *From Persepolis to the Punjab: Exploring Ancient Iran, Afghanistan and Pakistan*, edited by Elizabeth Errington and Vesta Curtis, 227–34. London: British Museum Press, 2007.

Wang Huijuan and Nemanja Cabric. "Interview: Belt and Road Initiative to Foster China-Serbia Win-Win Cooperation: Serbian Expert." *Xinhuanet*, August 14, 2016. http://news.xinhuanet.com/english/2016–08/14/c_135596795.htm.

Wang Kaihao and Hu Meidong. "Arts Festival Reflects Ancient Silk Road Cultures." State Council—People's Republic of China. November 9, 2015.

Wang, Wilfried. *Culture: City*. Baden, Germany: Lars Müller, 2013.

Wang Yi. "Toward a New Type of International Win-Win Cooperation." Speech at Luncheon of the China Development Forum, Beijing, March 25, 2015. http://www.fmprc.gov.cn/mfa_eng/zxxx_662805/t1248487.shtml?utm_campaign=buffer&utm_content=buffer1e863&utm_medium=social&utm_source=twitter.com.

War Memoryscapes in Asia Project. "What Is Warmap?" http://www.warinasia.com.

Warmington, Eric Herbert. *The Commerce between the Roman Empire and India*. Delhi: Vikas Publishing House, 1928.

Wastnidge, Edward. "Strategic Narratives and Iranian Foreign Policy into the Rouhani Era." *E-International Relations*. March 10, 2016. https://www.e-ir.info/2016/03/10/strategic-narratives-and-iranian-foreign-policy-into-the-rouhani-era/.

Waugh, Daniel. "The Making of Chinese Central Asia." *Central Asia Survey* 26, no. 2 (2007): 235–50.

———. "Richthofen's 'Silk Roads': Towards the Archaeology of a Concept." *Silk Road* 5, no. 1 (2007): 1–8.

———. "The Silk Roads in History." *Expedition* 52, no. 3 (2010): 9–22.

Weidong Sun. "Peaceful Development and Win-Win Cooperation." Speech at National Defence University, Islamabad, Pakistan, September 12, 2013. http://pk.china-embassy.org/eng/zbgx/t1076545.htm.

Weiss, Thomas. *Global Governance: Why? What? Whither?* Cambridge, UK: Polity, 2013.

Wekesa, Bob. "Admiral Zheng He and the Diplomatic Value of China's Ancient East African Contacts." Africa-China Reporting Project. July 9, 2013. http://china-africa-reporting.co.za/2013/07/admiral-zheng-he-and-the-diplomatic-value-of-chinas-ancient-east-african-contacts/.

Wen Jiabao. "Chinese Premier Wen Jiabao's Speech at the Arab League." Speech at the Arab League, Cairo, November 7, 2009. http://en.people.cn/90001/90776/90883/6806808.html.

———. "See China in the Light of Her Development." Speech at University of Cambridge, Cambridge, February 2, 2009. http://news.xinhuanet.com/english/2009-02/03/content_10753336_2.htm.

———. "Strengthen Good-Neighbourly Relations and Deepen Mutually Beneficial Cooperation." Speech at Balai Kartini, Indonesia, April 30, 2011. http://news.xinhuanet.com/english2010/china/2011-05/01/c_13853424.htm.

Wheeler, W. E. "The Control of Land Routes: Russian Railways in Central Asia." *Journal of the Royal Central Asian Society* 21, no. 4 (1934): 585–608.

Whitfield, Susan. *Life along the Silk Road.* Berkeley: University of California Press, 2015. Kindle.

———. "Was There a Silk Road?" *Asian Medicine* 3 (2007): 201–13.

Williams, Tim. *The Silk Roads: An ICOMOS Thematic Study.* Paris: International Council of Monuments and Sites, 2014.

Wilson, J. Keith, and Michael Flecker. "Dating the Belitung Shipwreck." In *Shipwrecked: Tang Treasures and Monsoon Winds,* edited by Regina Krahl and Alison Effeny, 35–37. Washington, DC: Arthur M. Sackler Gallery, Smithsonian Institution, and National Heritage Board, Singapore Tourism Board, 2010.

Winter, Tim. "Heritage Conservation Futures in an Age of Shifting Global Power." *Journal of Social Archaeology* 14, no. 3 (2014): 319–39.

———. "Heritage Diplomacy." *International Journal of Heritage Studies* 21, no. 10 (2015): 997–1015.

———. "Heritage Diplomacy: Entangled Materialities of International Relations." *Future Anterior* 21, no. 1 (2016): 16–34.

———. *Post-Conflict Heritage, Postcolonial Tourism: Culture, Politics and Development at Angkor.* London: Routledge, 2007.

———. *The Silk Road: A Biography.* Forthcoming.

Wong, Laura. "Relocating East and West: UNESCO's Major Project on the Mutual Appreciation of Eastern and Western Cultural Values." *Journal of World History* 19, no. 3 (2008): 349–74.

Wood, Frances. *The Silk Road: Two Thousand Years in the Heart of Asia.* Berkeley: University of California Press, 2002.

Wood, Robert. *The Ruins of Palmyra, Otherwise Tedmore in the Desart.* London: Robert Wood, 1753.

World Economic Forum. "Asia's Era of Infrastructure." January 22, 2016. https://www.weforum.org/events/world-economic-forum-annual-meeting-2016/sessions/asia-s-era-of-infrastructure/.

Wu Jianmin. "'One Belt and One Road,' Far-Reaching Initiative." *China-US Focus,* March 26, 2015. http://www.chinausfocus.com/finance-economy/one-belt-and-one-road-far-reaching-initiative/.

Wu Xiaogang, and Xi Song. "Ethnic Stratification amid China's Economic Transition: Evidence from the Xinjiang Uyghur autonomous region." *Social Science Research* 44 (2017): 158–72.

Xi Jinping. "Forging a Strong Partnership to Enhance Prosperity in Asia." Speech at the Singapore Lecture at the National University of Singapore, Singapore, November 7, 2015. http://learning.sohu.com/20151113/n426405360.shtml.

———. "A New Blueprint for Global Economic Growth." Speech at G20 summit, September 4, 2016. http://www.globalresearch.ca/chinas-president-xi-jinpings-opening-address-of-g20-summit-a-new-blueprint-for-global-economic-growth/5543895.

———. "President Xi's Speech to Davos in Full." Speech at World Economic Forum, Davos, January 17, 2017. https://www.weforum.org/agenda/2017/01/full-text-of-xi-jinping-keynote-at-the-world-economic-forum.

———. "Promote Friendship between Our People and Work Together to Build a Bright Future." Speech at Nazabayev University, Astana, September 7, 2013. http://www.fmprc.gov.cn/mfa_eng/wjdt_665385/zyjh_665391/t1078088.shtml.

———. "Speech by Chinese President Xi Jinping to Indonesian Parliament." Speech at Indonesian Parliament, Jakarta, October 2, 2013. http://www.asean-china-center.org/english/2013-10/03/c_133062675.htm.

———. "Speech by H.E. Xi Jinping President of the People's Republic of China at UNESCO Headquarters." Speech at UNESCO Headquarters, Paris, March 27, 2014. http://rs.chineseembassy.org/eng/xwdt/t1142560.htm.

———. "Work Together for a Bright Future of China-Iran Relations." Speech to Iranian newspaper, Tehran, January 21, 2016. http://www.fmprc.gov.cn/mfa_eng/wjdt_665385/zyjh_665391/t1334040.shtml.

Xi'an Jiaotong University. "Collaborative Innovation Centre of Silk Road Economic Belt Research Started by XJTU." Xi'an Jiaotong University, January 23, 2015. http://en.xjtu.edu.cn/info/1044/1572.htm.

Xie Feng. "Hand in Hand for Common Development." Speech at Indonesia's Ministry of Foreign Affairs, Indonesia, September 22, 2015. http://id.china-embassy.org/eng/gdxw/t1298984.htm.

———. "Partnership to Scale New Heights." *China Daily Asia*, April 17, 2015. http://www.chinadailyasia.com/asiaweekly/2015–04/17/content_15253195.html.

Xinhua News. "Cambodia Launches China-Backed Maritime Silk Road Research Centre." June 13, 2016. http://news.xinhuanet.com/english/2016–06/13/c_135432581.htm.

Yang Jiechi. "Jointly Build the 21st Century Maritime Silk Road by Deepening Mutual Trust and Enhancing Connectivity." Speech given at the Session of "Jointly Building the 21st Century Maritime Silk Road," Beijing, March 28, 2015. http://www.fmprc.gov.cn/mfa_eng/zxxx_662805/t1249761.shtml.

———. "Jointly Undertake the Great Initiatives with Confidence and Mutual Trust." Speech at the Boao Forum for Asia Annual Conference, Hainan, April 10, 2014. http://www.fmprc.gov.cn/mfa_eng/wjdt_665385/zyjh_665391/t1145860.shtml.

Yan Xuetong. *Ancient Chinese Thought, Modern Chinese Power.* Edited by Daniel Bell and Sun Zhe. Princeton, NJ: Princeton University Press, 2011. Kindle.

———. "From Keeping a Low Profile to Striving for Achievement." *Chinese Journal of International Politics* 7, no. 2 (2014): 153–84.

Yao Jianing, ed. "PLA Navy Training Ship Zhenghe Embarks on Far-Sea Voyage." *China Military,* October 25, 2016. http://english.chinamil.com.cn/view/2016-10/25/content_7324566.htm.

Yates, Donna, Simon Mackenzie, and Emiline Smith. "The Cultural Capitalists: Notes on the Ongoing Reconfiguration of Trafficking Culture in Asia." *Crime Media Culture* 13, no. 2 (2017): 245–54.

Yeh, Emily. "Introduction: The Geoeconomics and Geopolitics of Chinese Development and Investment in Asia." *Eurasian Geography and Economics* 57, no. 3 (2016): 275–85.

Yeh, Emily, and Elizabeth Wharton. "Going West and Going Out: Discourses, Migrants, and Models in Chinese Development." *Eurasian Geography and Economics* 57, no. 3 (2016): 286–315.

Ying Liu, Zhongping Chen, and Gregory Blue. *Zheng He's Maritime Voyages (1405–1433) and China's Relations with the Indian Ocean World: A Multilingual Bibliography.* Leiden: Brill, 2014.

York, Geoffrey. "Africa's Ambitious Mega-Port Project Shrouded in Skepticism." *Globe and Mail,* September 10, 2012. http://www.theglobeandmail.com/news/world/worldview/africas-ambitious-mega-port-project-shrouded-in-skepticism/article551398/.

Yoshihara, Toshi. "China's 'Soft' Naval Power in the Indian Ocean." *Pacific Focus* 25, no. 1 (2010): 59–88.

Yu Quiyu. *A Bittersweet Journey through Culture.* New York: CN Times Books, 2015. Kindle.

Zhao Hong. "The South China Sea Dispute and China-ASEAN Relations." *Asian Affairs* 44, no. 1 (2013): 27–43.

Zhang Dejiang. "Keynote Speech of China's Top Legislator at Belt and Road Summit in Hong Kong." Speech at Belt and Road Summit, Hong Kong, May 18, 2016. http://news.xinhuanet.com/english/china/2016-05/18/c_135368795.htm.

Zhang Weiwei. *The China Horizon: Glory and Dream of a Civilizational State.* Hackensack, NJ: World Century Publishing, 2016. Kindle.

———. *The China Wave: The Rise of a Civilisational State.* Hackensack, NJ: World Century, 2012.

Zhao Tingyang. *The Tianxia System: An Introduction to the Philosophy of a World Institution.* Nanjing: Jiangsu Jiaoyu Chubanshe, 2005. (Published in Chinese).

Zhou, Viola. "China Wins US$3 Billion Bid to Build Rail Line in Bangladesh." *South China Morning Post,* August 9, 2016. http://www.scmp.com/business/companies/article/2001471/china-wins-us3-billion-bid-build-rail-line-bangladesh.

Zhu Zhiqun. *China's New Diplomacy: Rationale, Strategies and Significance*. Farnham, UK: Ashgate, 2013. Kindle.

Żuchowska, Marta. "From China to Palmyra: The Value of Silk." In *Światowit: Annual of the Institute of Archaeology of the University of Warsaw*, 133–54. Warsaw: Institute of Archaeology of the University of Warsaw, 2013.

———. "'Grape Picking' Silk from Palmyra: A Han Dynasty Chinese Textile with a Hellenistic Decoration Motif." In *Annual of the Institute of Archaeology of the University of Warsaw*, edited by Franciszek Stępniowski, Andrzej Macialowiz, Ludwika Jończyk, and Dariusz Szelag, 143–62. Warsaw: Institute of Archaeology of the University of Warsaw, 2015.

Index

Locators in *italic* indicate figures. Endnotes are indicated by locators in the form "199n5," where the page number is given first, followed by the note number.